Bigelow Aerospace

Colonizing Space One Module at a Time

W0227485

Other Springer-Praxis books of related interest by Erik Seedhouse

Tourists in Space: A Practical Guide
2008
ISBN: 978-0-387-74643-2

Lunar Outpost: The Challenges of Establishing a Human Settlement on the Moon
2008
ISBN: 978-0-387-09746-6

Martian Outpost: The Challenges of Establishing a Human Settlement on Mars
2009
ISBN: 978-0-387-98190-1

The New Space Race: China vs. the United States
2009
ISBN: 978-1-4419-0879-7

Prepare for Launch: The Astronaut Training Process
2010
ISBN: 978-1-4419-1349-4

Ocean Outpost: The Future of Humans Living Underwater
2010
ISBN: 978-1-4419-6356-7

Trailblazing Medicine: Sustaining Explorers During Interplanetary Missions
2011
ISBN: 978-1-4419-7828-8

Interplanetary Outpost: The Human and Technological Challenges of Exploring the Outer Planets
2012
ISBN: 978-1-4419-9747-0

Astronauts for Hire: The Emergence of a Commercial Astronaut Corps
2012
ISBN: 978-1-4614-0519-1

Pulling G: Human Responses to High and Low Gravity
2013
ISBN: 978-1-4614-3029-2

SpaceX: Making Commercial Spaceflight a Reality
2013
ISBN: 978-1-4614-5513-4

Suborbital: Industry at the Edge of Space
2014
ISBN: 978-3-319-03484-3

Tourists in Space: A Practical Guide, Second Edition
2014
ISBN: 978-3-319-05037-9

Erik Seedhouse

Bigelow Aerospace

Colonizing Space One Module at a Time

 Springer

Published in association with
 **Praxis Publishing**
Chichester, UK

Dr Erik Seedhouse, M.Med.Sc., Ph.D., FBIS
American Astronautics Institute & Suborbital Training
Sandefjord, Norway

SPRINGER-PRAXIS BOOKS IN SPACE EXPLORATION

ISBN 978-3-319-05196-3   ISBN 978-3-319-05197-0 (eBook)
DOI 10.1007/978-3-319-05197-0
Springer Cham Heidelberg New York Dordrecht London

Library of Congress Control Number: 2014947104

Cover design: Jim Wilkie
Project copy editor: Christine Cressy
Front cover image courtesy: Boeing
Back cover images courtesy: Bill Ingalls/NASA

Printed on acid-free paper

Springer is part of Springer Science+Business Media (www.springer.com)

Contents

Acknowledgments

In writing this book, the author has been fortunate to have had five reviewers who made such positive comments concerning the content of this publication. He is also grateful to Maury Solomon at Springer and to Clive Horwood and his team at Praxis for guiding this book through the publication process. The author also gratefully acknowledges all those who gave permission to use many of the images in this book, especially Bill Ingalls of NASA, Paul Lemon, and Mary Kane at Boeing, for providing the cover image of this book.

The author also expresses his deep appreciation to Christine Cressy, whose attention to detail and patience greatly facilitated the publication of this book, and to Jim Wilkie for creating yet another eye-catching cover, and to D. Raja and Hemalatha Gunasekaran for their meticulous attention in bringing this book to publication.

To
Drs. David Grundy and Paul Enck,
for their invaluable guidance and support
of my academic pursuits

About the Author

Erik Seedhouse is a Norwegian-Canadian suborbital astronaut whose life-long ambition is to work in space. After completing a degree in Sports Science at Northumbria University, he joined the legendary 2nd Battalion the Parachute Regiment. During his time in 2-Para, Erik spent six months in Belize, training in the art of jungle warfare. Later, he spent several months learning the intricacies of desert warfare in Cyprus. He made more than 30 jumps from a C130, performed more than 200 helicopter abseils, and fired more anti-tank weapons than he cares to remember!

Upon returning to the comparatively mundane world of academia, the author embarked upon a master's degree in Medical Science at Sheffield University. He supported his studies by winning prize money in 100-km running races. After placing third in the World 100-km Championships in 1992 and setting the North American 100-km record, the author turned to ultra-distance triathlon, winning the World Endurance Triathlon Championships in 1995 and 1996. For good measure, he won the inaugural World Double Ironman Championships and the Decatriathlon, a diabolical event requiring competitors to swim 38 km, cycle 1,800 km, and run 422 km. Non-stop!

Returning to academia in 1996, Erik pursued his Ph.D. at the German Space Agency's Institute for Space Medicine. While studying, he found time to win Ultraman Hawai'i and the European Ultraman Championships as well as completing Race Across America. As the world's leading ultra-distance triathlete, Erik was featured in dozens of magazines and television interviews. In 1997, *GQ* magazine nominated him the "Fittest Man in the World".

In 1999, Erik retired from being a professional triathlete and started post-doctoral studies at Simon Fraser University. In 2005, he worked as an astronaut training consultant for Bigelow Aerospace and wrote *Tourists in Space*. He is a Fellow of the British Interplanetary Society and a member of the Space Medical Association. In 2009, he was one of the final 30 candidates in the Canadian Space Agency's Astronaut Recruitment Campaign. Erik works as an astronaut instructor with the American Astronautics Institute, triathlon coach, and author. He is Editor-in-Chief of the *Handbook of Life Support Systems for Spacecraft*, a major reference work to be published by Springer in 2016, and is the Training Director for Astronauts for Hire (www.astronauts4hire.org). Between 2008 and 2013, he was director of Canada's manned centrifuge operations.

In addition to being a suborbital astronaut, triathlete, centrifuge operator, pilot, and author, Erik is an avid mountaineer and is pursuing his goal of climbing the Seven Summits. *Bigelow Aerospace* is his 15th book. When not writing, he spends as much time as possible in Kona and at his home in Sandefjord. Erik and his wife, Doina, are owned by three rambunctious cats—Jasper, Mini-Mach, and Lava.

Author (left) holding the Subida Veleta 50 kilometer trophy

Acronyms

ACS	Attitude Control System
AES	Advanced Equipment System
AMIS	Advanced Man In Space
ARRA	American Recovery and Reinvestment Act
ATV	Automated Transfer Vehicle
BAASS	Bigelow Aerospace Advanced Space Studies
BEAM	Bigelow Expandable Activity Module
CAM	Centrifuge Accommodation Module
CAN	Controller Area Network
CBM	Common Berthing Module
CCDev	Commercial Crew Development
CCiCap	Commercial Crew Integrated Capability
CCP	Commercial Crew Program
CDR	Critical Design Review
CHeCS	Crew Health Care System
CMS	Case Management System
COTS	Commercial Orbital Transportation System
CQ	Crew Quarters
CRS	Crew Resupply Services
DTC	Defence Trade Control
DTSA	Defense Technology Security Administration
ECLSS	Environmental Controlled Life-Support System
ESA	European Space Agency
EVA	Extravehicular Activity
FAA	Federal Aviation Administration
GTO	Geostationary Transfer Orbit
HEDS	Human Exploration and Development of Space
HSP	Habitation System Project
ICBM	Intercontinental Ballistic Missile
IRDT	Inflatable Re-entry Development Technology

ISF	Industrial Space Facility
ISPCS	International Symposium for Personal and Commercial Spaceflight
ISS	International Space Station
ITAR	International Trade on Arms Regulations
JPL	Jet Propulsion Laboratory
JSC	Johnson Space Center
LBSS	Lunar Base Systems Studies
LEO	Low Earth Orbit
MISSE	Materials on International Space Station Equipment
MMOD	Micrometeoroid Orbital Debris
MMSEV	Multi Mission Space Exploration Vehicle
MUFON	Mutual UFO Network
NACA	National Advisory Committee on Aeronautics
NAUTILUS	Non-Atmospheric Universal Transport Intended for Lengthy US Xploration
NEO	Near Earth Object
NIDS	National Institute for Discovery Science
OSC	Orbital Sciences Corporation
OST	Outer Space Treaty
PBO	Poly Phenylene
PCBM	Passive Common Berthing Mechanism
PET	Polyethylene Terephthalate
PVC	Polyvinyl Chloride
SAA	Space Act Agreement
SDU	Shell Development Unit
SEIM	Surface Endoskeletal Inflatable Module
SLS	Space Launch System
SMAC	Spacecraft Maximum Allowed Concentration
TAAT	Technology Applications Assessment Team
UHMWPE	Ultra High-Molecular-Weight Polyethylene
ULA	United Launch Alliance
WSTF	White Sands Test Facility

Author Note

Since the founding of Bigelow Aerospace, there has always been a degree of mystique surrounding the company. While located in Las Vegas, a city known for flashy spectacle and over exuberance, Bigelow Aerospace has kept a low profile. A very low profile. For many years, people in the spaceflight industry knew Bigelow was working on something that involved the inflatable habitat technology, but there was little publicity. Shying from media attention, the lone-wolf visionary refused to have photos taken by or for the press, and denied television interview requests. The shyness was interpreted by some in the media as a sign of secrecy, and (completely unsubstantiated) claims were made that Bigelow was associated with intelligence agencies. Given the company's secret-squirrel profile and given that the company founder owned the Budget Suites of America hotel chain, it was inevitable that some media outlets suggested Bigelow was in the business of building space hotels. The reality, as you will read in this book, is very different. One of the challenges of writing a book about a company that maintains such a low profile is access to information. You will notice that none of the images in this book is a Bigelow image. Requests were made to the company asking permission to use images, as were fact-checking requests, but no response was received, which means the book you are reading is an independent account of Bigelow Aerospace.

Foreword

"NASA Awards $17.8 Million for an Inflatable Addition to the ISS"

That was the headline posted in January 2013 after Bigelow Aerospace landed its first major deal with NASA and a chance to prove that the future of human space exploration is inflatable (the company prefers the term "expandable"). Whilst SpaceX might be the toast of the commercial spaceflight—NewSpace—industry thanks to their bold achievements and lofty goals, Elon Musk's operation is far from the only player in the game.

While SpaceX is in the business of getting cargo and crew into orbit, keeping them there is someone else's job, and that someone happens to be Bigelow Aerospace. Unlike most NewSpace companies that are focusing on the launch side of the equation, Bigelow is focusing solely on the "staying up there" part. To that end, he is developing technology for a revolutionary kind of space station that promises to deliver much larger volumes at a fraction of the cost of traditional space station modules. NASA's 2013 contract with Bigelow is proof that the nascent space company is now at the point where its technology is ready for prime time.

Bigelow Aerospace has been trying to get the world to take its expandable space habitats seriously for years and, while some have regarded the Vegas-based firm's habitats with skepticism, NASA has always been willing to listen to Bigelow's ideas. And now the space agency is investing in them. NASA has awarded the private space contractor a US$17.8 million contract to develop a new addition to be berthed with the International Space Station (ISS) in 2015. This is big news for NASA and Bigelow, who has been working for years to create space habitats for corporate clients and government space agencies, going so far as to propose concepts for inflatable Moon bases. The company has already launched two orbiting prototypes—*Genesis I* and *Genesis II*—but, until the NASA deal came along, no government or corporate entity had bought Bigelow's technology. NASA, for its part, is looking for inexpensive and proven ways to expand the space aboard the ISS, and a Bigelow Expandable Activity Module (BEAM) could do exactly that.

Here for the first time, in *Bigelow Aerospace*, you can read how a space technology start-up company is pioneering work on expandable space station modules. Here you will learn how Bigelow Aerospace was founded and how it is funded in large part by the fortune Bigelow amassed through his ownership of Budget Suites of America. This book explains how Bigelow originally licensed the multilayer, expandable space module technology from NASA after Congress canceled the ISS TransHab project following delays and budget constraints. It explains how the entrepreneur continued to develop the technology for a decade, redesigning the module's fabric layers—including adding proprietary extensions of Vectran shield fabric, "a double-strength variant of Kevlar"—and developing a family of uncrewed and crewed expandable spacecraft in a variety of sizes. Here, in *Bigelow Aerospace*, you can read how NASA came full circle to once again consider connecting a Bigelow expandable craft to the ISS for safety, life support, radiation shielding, thermal control, and communications verification testing. Also described is Bigelow's unique business model that involves leasing out voluminous space stations to research communities or corporations—in October 2010, Bigelow announced it had agreements with six sovereign nations to utilize on-orbit facilities of its commercial space station: the UK, The Netherlands, Australia, Singapore, Japan, and Sweden. And at the core of the plan are the inflatable modules. In this book, you can read how these modules are actually tougher and more durable than rigid modules, thanks to the company's use of several layers of Vectran, a material twice as strong as Kevlar, and also because, in theory, flexible walls are able to sustain micrometeoroid impacts better than rigid walls.

Also described is how the link between Bigelow Aerospace, NASA, and dozens of private companies can lead to the creation of a new economy—a space economy. It describes the wait for launch capabilities and Bigelow's plans for a commercial space station—the Bigelow Next-Generation Commercial Space Station, a private orbital space complex under development. Finally, Chapter 8 describes the company's aspirations beyond LEO and how Bigelow aims to look at ways for private ventures to contribute to human exploration missions, some of which may include establishing bases on the Moon and beyond.

1

Robert T. Bigelow, Orbital Real Estate Mogul

There are some who label Mr. Bigelow (no one calls him by his first name) an eccentric. After all, this is the guy who once gave an estimated US$10 million to fund the unidentified flying object (UFO)-hunting National Institute for Discovery Science (NIDS). He also bought a 480-acre cattle ranch in Utah that some believe is the site of an inter-dimensional doorway used by alien shape-shifters—a subject we'll get to shortly. But, eccentric or not, what cannot be denied is that Bigelow is rich. Very rich. While Bigelow has never been on the Forbes 400 richest people list, Forbes estimates the Las Vegas native's real estate empire is worth US$700 million, thanks to the Budget Suites chain of residential hotels and more than 14,000 apartment and office units Bigelow owns across the South-West. Today, Bigelow (Figure 1.1), who made his fortune building affordable places to stay on Earth, is looking to do the same in space—by building high-tech, low-cost inflatable space stations. In the last 15 years, he has spent US$210 million of his personal fortune and says he will spend up to US$500 million to prove that space is a safe place for entrepreneurs.

BIGELOW: FROM BUDGET SUITES TO BEAM

Bigelow grew up in Las Vegas, a town as unique as the self-made millionaire. In the 1940s and 50s, Vegas was a place where it was possible to witness rocket launches and even the occasional nuclear weapons test. It was also a place where strange events occurred, which perhaps isn't so surprising given that Sin City is only a two-and-a-half-hour drive from the mysterious Area 51. One such event took place in 1947 when Bigelow's grandparents were driving across Mount Charleston near Las Vegas and witnessed a UFO approach their car. When the young Bigelow heard the tale, it sparked an interest in the paranormal (he's never had a close encounter) and space, although Bigelow never aspired to become an astronaut and he never wanted to work for NASA. Bigelow's dream was to explore the universe on his own terms, which meant he needed to make money, and a lot of it. Bigelow's father was a broker, so real estate seemed a viable route to wealth. To that end, Bigelow studied real estate and banking at Arizona State University, graduating in 1967. Then he got to work. He began by borrowing US$20,000, using the money to buy small rental apartment complexes. Three years later, he owned about 100 apartments in Las Vegas and had begun work on a 40-unit

© Springer International Publishing Switzerland 2015
E. Seedhouse, *Bigelow Aerospace: Colonizing Space One Module at a Time*,
Springer Praxis Books, DOI 10.1007/978-3-319-05197-0_1

1.1   Robert T. Bigelow. Courtesy: Bill Ingalls/NASA

apartment building. He was only 26 years old. The apartment building was followed by several more, which eventually became a chain—Budget Suites of America. Founded in 1988, Budget Suites of America is a chain of extended-stay hotels offering apartments for rent by the week. The convenience and affordability of a Budget Suites hotel mean they're popular with the temporary or migrant workers who work in The Strip's hotels and casinos. In addition to making enough money to establish Bigelow Aerospace (www.bigelowaerospace. com), the Budget Suites business model provided a template for Bigelow's space stations that would also be designed to be cheap, efficient, and available for monthly lease.

I met Bigelow in 2005 in Las Vegas to talk about astronaut training. He had arranged for us to have lunch at Marie Callenders (Bigelow is a fan of their cheesecakes) on Rainbow Boulevard. Bigelow arrived in a black SUV, a tall, trim 60-year-old with a full head of silvery-black hair wearing crisply pressed beige slacks and a short-sleeved dress shirt. Over the next hour, he talked about the challenges of the International Trade on Arms Regulations (ITAR) and his strategy for overcoming the lack of affordable transportation options for getting humans into space and to his habitats. To say the guy was knowledgeable about manned spaceflight and aerospace technology is an understatement—I could have been talking to a NASA engineer. He was very clear about the elements needed to make his aerospace dreams a reality—a master plan that will mass produce inflatable space stations and, in so doing, revolutionize human access to space. Since our meeting, his company has gone from strength to strength and one of his BEAM (Bigelow Expandable Activity Module) habitats is due to be attached to the International Space Station (ISS) in 2015. Way back when Bigelow started talking about inflatable habitats, there were many who thought the scheme preposterous. But for Bigelow, born, raised, and still living in the world's most preposterous city, such a goal was anything but. After all, Vegas was built on the proposition that, if you've got the money, you can create anything you want. Recreate Venice, complete with indoor canals? No problem. A beach with a wave machine so tourists can body surf in the middle of the desert? Done. A privately owned, inflatable space station? Why not?

THE UFO CONNECTION

"I have a huge concern how the American people are going to react to the first contact. How many people are going to go to the gun shop? How many are not going to go to work?"

Robert Bigelow

Those of you who follow Bigelow Aerospace will no doubt have noticed the company's logo—an iconic large-eyed alien that is also prominent on the company's Mission Control screens. To those in the UFO community, Bigelow is a familiar name, having doled out millions of dollars to fund research into alien abductions and UFO sightings. His interest in the phenomenon can be traced back to the experience related to him by his grandparents who witnessed a UFO. At first they thought it was an airplane on fire, but it was moving much faster than an airplane, and its light filled their windshield before making a right-angled turn and shooting off into the sky. Bigelow first heard this tale when he was 10. There are some who suggest one of the reasons he's so eager to start orbiting his space stations is to get one step closer to making contact. There are others who claim he fears that UFOs may interfere with his space stations and yet others who contend Bigelow believes that UFOs may harbor secrets of propulsion that his engineers might be able to put to good use! Whichever rumors you choose to believe, the fact is that Bigelow has been engaged with the UFO community for some time. For example, he was the principal sponsor of the Las Vegas-based NIDS from its founding in 1995 until it was placed on inactive status in 2004—the NIDS staff included several PhDs and ex-FBI agents who researched alien abductions, out-of-body experiences, cattle mutilations, and other paranormal phenomena. The NIDS website is still up (www.nidsci.org) but has not been updated since 2004. It reports UFO investigations, alleged cattle mutilations, and other extraterrestrial-related events. Perhaps the most controversial project undertaken by NIDS was its purchase of a ranch in Utah, which some describe as a hyper-dimensional portal. The ranch is said to be infested by an alien creature known as Skinwalker, taking its name from Native American legends similar to European legends about werewolves. NIDS researchers investigating the ranch with sophisticated electronic equipment failed to obtain any actual proof that anything unexplainable was going on. Bigelow has also met with more than 200 people who claim to have witnessed aliens and, in 1997, he donated US$3.7 million to the University of Nevada, Las Vegas, to create a Consciousness Studies program, which offered classes about near-death experiences and psychic phenomena.

Not surprisingly, Bigelow's involvement in the UFO community has resulted in some labeling him an eccentric, but Bigelow stresses the research has been carried out with scientific rigor and most of the people he has interviewed have jobs in the military or the sciences. Having invested so much time and money in the UFO enigma, it isn't surprising that Bigelow has compiled a sizeable amount of data from a lot of different sources that give him some pretty strong convictions about the existence of UFOs, as evidenced by this sound-bite posted on the NIDS website in 2000:

"I strongly believe that at least *some* UFOs owe their beginnings to being manufactured … from materials made in a microgravity environment. As for our UFO friends, we will not begin to match their early craft until we also begin to exploit space for manufacturing purposes."

Another interesting element linking Bigelow with the UFO community is a strange Federal Aviation Administration (FAA) directive (see below) issued in 2010 showing that the US Government deferred all FAA UFO reports to Bigelow Aerospace.[1]

U.S. DEPARTMENT OF TRANSPORTATION
FEDERAL AVIATION ADMINISTRATION

ORDER
JO 7110.65U

Air Traffic Organization Policy

Effective Date:
February 9,
2012

Subject: Air Traffic Control

Includes: Change 1 effective 7/26/12, Errata to Change 1 effective 7/26/12, Change 2 effective 3/7/13, and Errata to Change 2 effective 3/7/13. Change 3 effective 8/22/13.

Section 8. Unidentified Flying Object (UFO) Reports

9-8-1. GENERAL

a. Persons wanting to report UFO/unexplained phenomena activity should contact a UFO/ unexplained phenomena reporting data collection center, such as Bigelow Aerospace Advanced Space Studies (BAASS) (voice: 1-877-979-7444 or e-mail: Reporting@baass.org), the National UFO Reporting Center, etc.

---

[1] Although BAASS receives no funds for processing UFO reports, the collection center has hotline staff on duty to receive UFO reports.

MUFON

Not surprisingly, the directive raised eyebrows in the UFO-investigative community because, until the 2010 FAA directive, it had been the Mutual UFO Network (MUFON) that was the go-to organization for UFO investigations. MUFON, an American non-profit organization, is one of the largest UFO-investigative organizations in the world, with more than 3,000 members, including 800 field investigators. The organization operates a world-wide network of regional directors for field investigation of reported UFO sightings and holds an annual international symposium. What many in the UFO community were wondering was how the Bigelow connection tied in with the FAA and MUFON and how BAASS tied in with MUFON. The result was a hybrid partnership between for-profit BAASS and non-profit, volunteer-driven MUFON, and the setting-up of a Star Team Impact Project (SIP Project), a union that paid investigators engaged in cases where the physical effects of a UFO were reported or where "living beings" were sighted. Anyone who was already a MUFON investigator could apply for a position with the project. Although the arrangement seemed an interesting model to investigate UFOs, a clash of corporate and non-profit cultures and lack of mutual communication framework between an accounting-oriented corporate contractor and a mission-driven volunteer organization triggered a suspension and review of the agreement.

The Star Impact Project

To understand why the SIP failed, it's necessary to delve into the details. The SIP was a MUFON program funded in part by BAASS where MUFON was subcontracted to provide information from the Case Management System (CMS) database and witness reports directly to BAASS. MUFON had a contract with BAASS requiring MUFON to provide data from reports that were submitted to its CMS website in exchange for BAASS paying funds to MUFON each month. Part of the money was used to pay investigators deployed to investigate UFO cases and members monitoring the cases coming into the CMS website. To begin with, the arrangement ran smoothly. MUFON provided data to BAASS about events that were reported in CMS and, upon completion of every SIP Project Investigation, a complete investigative report was shared with BAASS. Every Friday, MUFON and BAASS discussed the weekly reports, reviewed cases, and discussed issues between the two organizations. The SIP Project ran for several months, during which time 65 SIP cases were investigated. Then financial issues arose, which ultimately led to the project's demise.

The Contract Agreement required US$56,000 to be paid to MUFON each month. It was a one-year agreement initiated at the end of February 2009 which ran to the end of February 2010. MUFON received about US$324,000 from the BAASS SIP Project, which was a little less than half of the original contract deal that could have paid a total of US$672,000 in the first year. The first signs of trouble surfaced during the first of four-monthly contract performance reviews in June 2009 when concerns were raised by BAASS about excess funds that were not spent on the project by MUFON. Following the review, BAASS reduced the amount of money to US$25,000 a month. Not surprisingly, each party had opposing thoughts on the issue, with MUFON taking a position that they could keep the excess money, without accounting to BAASS for how they used it: at the time of the review, MUFON was spending about US$25,000 per month on the project and had accumulated about US$120,000 in savings, which was not spent on the SIP Project. Conversely, BAASS maintained they expected that most of funds should be spent on the SIP Project.

Despite animosity between the two organizations, the SIP Project was running well and BAASS was pleased with the case reports being processed. Behind the scenes, however, there was talk about BAASS trying to control MUFON—a situation which deteriorated when certain payments were not disclosed to BAASS. By the time the October review came around, MUFON had another problem to contend with because BAAS announced they were going to reduce funding to US$15,000, thereby forcing MUFON to spend some of the money in the savings. MUFON refused the deal and no funding was received from BAASS after that. It was a blow to MUFON, whose operating expenses ran to about US$28,000 per month. As the organization rapidly ran out of money, negotiations with BAASS intensified. MUFON's case wasn't helped when BAASS made several requests for more detailed financial information about how MUFON spent the funding, which were not acknowledged. Eventually, BAASS requested all of MUFON's bank statements from the SIP Project and the MUFON checking accounts. Discrepancies aroused the attention of BAASS auditors and the SIP Project shut down. MUFON was broke and had to close the office in Fort Collins and lay off the last of the two staff employees.

Skinwalker

Since the BAASS–MUFON collaboration, Bigelow has been the subject of more than a few conspiracy articles and documentaries, one of which was in Season 3 of the *Conspiracy Theory with Jesse Ventura* television series[2] hosted by Jesse Ventura, which focused on the events at Skinwalker Ranch. The episode promised to deal with one of the most far-out conspiracies yet—a cover-up of an extraterrestrial invasion. Named after the Skinwalker Ranch, Utah, with a history of paranormal phenomena and possible UFO interaction, "Skinwalker" is a term from Native American culture to describe supernatural entities that have the ability to shape-shift. The ranch was bought by Bigelow in 1996, after the previous owners were allegedly driven away from the property by paranormal events. According to the program's narrator, believers in the Skinwalker Conspiracy thought Bigelow had bought the ranch to investigate these paranormal phenomena, and that he might even have been in contact with aliens. To investigate the conspiracy, Ventura meets with Robert Regehr, a believer in the Skinwalker Conspiracy, and retired scientist who worked with NASA on the Apollo missions. Regehr explains that with NASA losing its heavy-launch capability, corporations are taking over to fill the void left by the agency. He goes on to explain that these private corporations will begin to do things that are constrained by international treaties, and alludes to the possibility they could be colluding with aliens!

From there, the program turns its focus on Bigelow and Don Ecker, former research director for *UFO Magazine*. Ecker explains that Bigelow and his team spent eight years searching for evidence of aliens at the ranch, but then mysteriously stopped in 2004, with all the data locked away. Next to be interviewed is nuclear physicist and ufologist

---

[2]   Episode 305 (20), "Skinwalker", aired on December 3rd, 2012.

Dr. Franklin Ruehl, who claims Bigelow knows something the rest of us don't, quoting an interview with Bigelow from the *New York Times* in June 2010, in which Bigelow, when asked about UFOs, stated that "People have been killed. People have been hurt". When asked later about this quote, Bigelow claims he was referring to the Colares, Brazil, incident from 1977, during which radiation beams fired by a UFO injured or killed up to 35 people. Next, the program's investigators meet with paranormal researcher, Christopher O'Brien, for further insight into Skinwalker Ranch. O'Brien contends that Bigelow's agenda is to collect as much scientific data as he can to give his company an advantage when designing advanced propulsion systems via reverse engineering.

Determined to learn more, the investigative team decide to go on a night-time stakeout at Skinwalker Ranch, where they spot a UFO—a bright light that hovers, makes a sharp turn, darts across the heavens, disappears for a moment, then reappears. The narrator notes that, according to the FAA directive of February 11th, 2010, all UFO reports should be forwarded to the BAASS Group. Jesse Ventura thinks this is strange and decides to meet Bigelow face to face. The film crew heads to the Bigelow Aerospace complex, where Ventura leads them to the main entrance, which is guarded by military types sporting alien logos on their hats. Jesse asks one of the guards if there are aliens on the property. The guard stonewalls, saying "I'm not able to answer any questions about what may or may not be on Mr. Bigelow's property. That would have to be routed through corporate, sir". Eventually, Jesse tracks down Bigelow at a conference in Las Cruces, New Mexico. Jesse spots Bigelow as he's coming out of a conference room, introduces himself as Governor Ventura, and asks: "I heard that whenever there's a UFO sighting, they defer to you, the government. Is that true?" "It's a passion I've had for a lot of years. I have a company that is referred in the FAA manuals, that we're supposed to receive a phone call," explains Bigelow. When Jesse asks how that happened, Bigelow replies "that's a long story". Bigelow's people hustle him away but, a few minutes later, one of the crew catches up with Bigelow and resumes the line of questioning. "Years ago, I had to make a decision that it was probably best that we not talk too much," explains Bigelow. When asked if he believes in UFOs, Bigelow's reply is as direct as they come: "Most folks have somebody in their family that has seen something very strange, that was anomalous. So, yes, I do believe that there is enough evidence, enough proof, that a prosaic explanation doesn't cover the phenomenon."

The next expert interviewed is Ventura's friend, conspiracy theorist and alternative media mogul Alex Jones, who reckons Bigelow is an unwitting front man. "The guy who's really behind this is Col. John B. Alexander," Jones says. The narrator explains that Alexander was part of a military project that tried to develop a super soldier who could walk through walls, which served as part of the inspiration for the film *The Men Who Stare at Goats*. According to the narrator, Alexander worked for Bigelow on his UFO hunt and also led a secret study into the government's UFO files, which concluded UFOs are real and controlled by superior intelligence. In a face-to-face meeting between Alexander and Ventura, Alexander denies he still works for Bigelow, but admits to setting up the UFO reporting. Asked about Skinwalker Ranch, Alexander is vague, saying only that "It's a very strange place". The response doesn't satisfy Ventura, who attempts to delve deeper, asking what sort of strange events happened at the ranch and whether the government was being truthful about what they have told the public about the existence of aliens. It's a question that irritates Alexander, who doesn't reveal anything.

The investigation concludes with a meeting with MUFON investigator Preston Dennett, who claims to have important information about Bigelow Aerospace and Skinwalker Ranch. According to Dennett, Bigelow knows what the aliens know, and reckons the business of launching habitats in space is so Bigelow can use it as an escape pod. "This is no joke," says Dennett. As with most episodes, the investigation ends with more questions than answers. Ventura sums up by saying: "A conspiracy of silence ain't the answer. And what's going on at that Skinwalker Ranch anyway? I'm Jesse Ventura, and this is Conspiracy Theory. Tune in next week for Episode #6: Brain Invaders."

BIGELOW AEROSPACE

TransHab

Bigelow didn't set out to start his own space business. In 1996, he invested in a couple of emerging commercial space companies but, once he witnessed the company's business sense—or lack of it—Bigelow declined the board seats offered to him and decided to go his own way. In 1999, he founded Bigelow Aerospace, although he wasn't quite sure what the company would be building. One option was to develop a spacecraft that could accommodate 100 passengers on a round-the-Moon voyage. Then, in 1999, Bigelow came across an *Air & Space*/Smithsonian article, "Launch. Inflate. Insert Crew", about a US$100 million NASA project called TransHab. TransHab was a lightweight inflatable habitat, made of tough, bullet-proof fabric, designed to shelter astronauts en route to Mars. At the time, the program, like so many space ventures, was in danger of being cut, but it wasn't the funding challenges that caught Bigelow's attention: TransHab (Figure 1.2) was touted as a possible habitat to be docked with the ISS.

With their lower weight and smaller volume, Bigelow reckoned the inflatables could be configured as space stations, providing habitats that would be much less expensive to launch. A few months later, after more Congressional scrutiny, NASA was forced to ditch TransHab and Bigelow began negotiations with the agency to license the technology under a Space Act Agreement (SAA). Three years later, Bigelow secured the rights to TransHab's eight patents. The only thing missing was an instruction manual! Undeterred, Bigelow recruited some of the brightest and best engineers from TRW, Boeing, and Raytheon, who went to work filling in the gaps. Unfortunately, knowledgeable as they were, Bigelow's team didn't have the experience of the NASA engineers. Every once in a while, the team would hit a dead end and the name "Schneider" would be mentioned. After this had happened a few times, Bigelow asked who this Schneider guy was. "Schneider" was William Schneider, a senior engineer at NASA and considered the father of TransHab. He had retired in 2000 and was working in the engineering faculty at Texas A&M University. Bigelow invited him to Las Vegas and, shortly afterwards, Schneider started work as a consultant at Bigelow Aerospace. With Schneider on board, the technical challenges were solved, but a bigger headache loomed: politics.

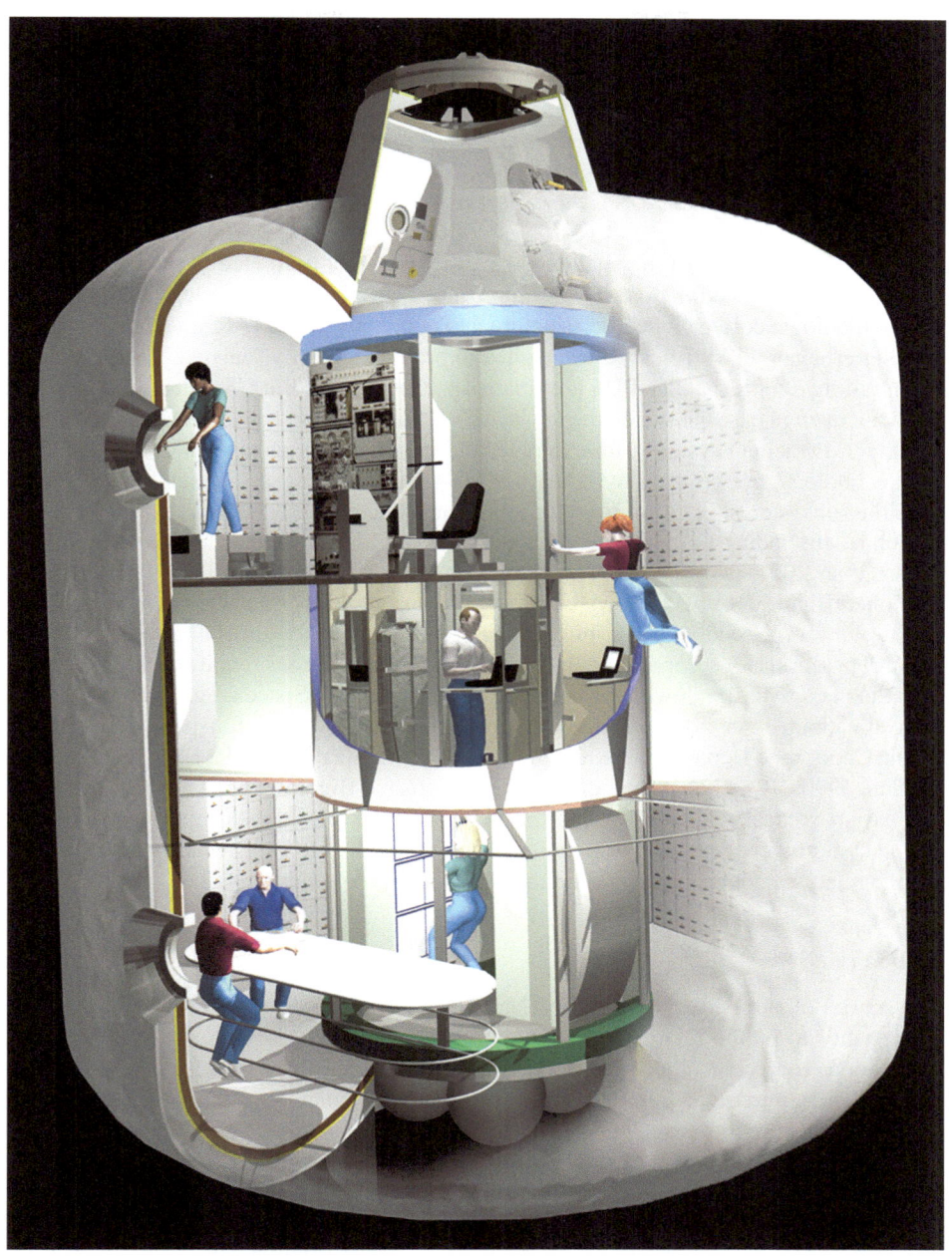

1.2 TransHab vertical interior cut-away view. Courtesy: NASA-JSC, Frassanito and Associates

Dealing with ITAR

While the technological challenges of any space endeavour are huge, engineering isn't at the top the list of things that can hurt a space enterprise. Politics is. In short, if the company doesn't have the political environment or political permission to conduct a particular activity, the technology is more or less worthless. What companies like Bigelow Aerospace are most concerned with in the US is the export-control regime, because the State Department has no incentive to promote and assist a smooth relationship between domestic and foreign partners. One example of the challenge faced by Bigelow presented itself when the company built a test stand for *Genesis I*, a simple metal structure with four legs that looked like an upside-down coffee table. The structure had to comply with ITAR, and that's where the problems began. ITAR rules were designed to protect militarily sensitive US technologies from falling into the hands of US adversaries such as China. But US allies are also subject to them, even in cases in which the law's application seems to have escaped the bounds of its intent. Parachute systems, for example, and more broadly any entry, descent and landing technology, are covered by "dual use" ITAR regulations, as are satellites and many satellite subsystems. For a company trying to launch a rocket, ITAR is a real can of worms, which is why industry behemoths such as Lockheed Martin have small armies of lawyers (more than 200—two hundred!) working just on the ITAR issue. The ITAR inspectors reckoned Bigelow's upside-down coffee table violated ITAR rules and so Bigelow had to get to work processing a paper blizzard of administration to get in line with the export-control process. Not surprisingly, Bigelow's view of the onerous requirements of ITAR are less than positive: Bigelow believes, as do many in the NewSpace arena, that the best way to deal with the issue is to build support for moving export control for space hardware back to the Commerce Department, where it resided before the move to the State Department in the late 1990s. It's a belief that, in December 2007, led Bigelow to file a legal challenge to the export-control regime disputing that foreign passengers travelling on a spaceship or space station were involved in a transfer of technology. He won, which proved a minor public relations coup, at least within the NewSpace community.

Public relations

The corps of entrepreneurial space (NewSpace) companies have taken very different approaches to public relations. At one extreme are companies such as Blue Origin that have adopted a secret squirrel approach (outside of a federally mandated environmental assessment report) that puts the National Security Agency to shame and, at the other extreme, there are those, such as Armadillo Aerospace, that have been very open about the status of their development efforts, sharing the latest news with the media and the public: Armadillo blogs almost all the details of their work, regardless of success or failure. For most of its early existence, Bigelow Aerospace was in the same camp as Blue Origin. In the first few years after its founding, the company shared few details of its work but, in recent years, the company has gradually warmed to the media and the public, especially following the successful launch of its first spacecraft, *Genesis I*. Following the *Genesis I* launch, the company has opened the door a little wider, as evidenced by Bigelow's speech at the International Symposium for Personal and Commercial Spaceflight (ISPCS) in Las Cruces, New Mexico, in October 2011.

SPACE AMBITIONS

Bigelow's ISPCS speech

Those attending the 2011 ISPCS thought perhaps Bigelow might share his company's space plans. But Bigelow spent very little time talking about his company's ambitions, choosing instead to spend most of his 40-minute speech focusing on his belief that, within 15 years, China will not only land people on the Moon, but will claim it as its own territory. Why would China do this? Bigelow believes China's ambition to be the most powerful country in the world will drive the nation to lay claim to the Moon. What this means, Bigelow continued, is not just repeating past achievements in space, but moving beyond them. "Why not take the all-important syllogistic next step: ownership, ownership, ownership?" he suggested. Doing so, he said, would generate "global psychological impact" and considerable prestige for the Chinese people. "I think nothing else the Chinese could possibly do in the next 15 years would cause as great a benefit for China," he continued. Bigelow argued that China, with its growing wealth, would be able to land humans on the Moon and start making claims as early as 2022. One obstacle to any Chinese lunar claims is the Outer Space Treaty (OST), of which China is a party. The OST prohibits countries from making territorial claims to the Moon, but Bigelow suggested China could amend the treaty through the support of countries where China is making investments. Alternatively, Bigelow suggested, China could simply withdraw from the treaty. Such an action would restore a "fear factor" kind of motivation for US space efforts that has been lacking since the fall of the Soviet Union. Bigelow suggested the US had weakened in recent years because of increasing entitlement spending and a lack of quality in the nation's education system. The fear instilled by China claiming the Moon, Bigelow argued, could stimulate the space program and the country in the same way as the space race with the Soviet Union did, although he was skeptical whether this would work: "It may, and probably will be, much too late for the Americans to respond to China securing the Moon. However, Mars will likely still be available."

Bigelow's ISPCS speech wasn't the first time he had talked about China making lunar claims (he gave a similar speech at the 14th Annual FAA Commercial Space Transportation Conference), but the ISPCS speech gained more media attention. Were Bigelow's concerns valid? It's difficult to say, because of the lack of official information about Chinese plans for *any* of its space missions, never mind human lunar missions. What we do know is China has announced that scientists interested in human lunar exploration have formed a working group to conduct a feasibility study for a human lunar mission. We also know they plan on constructing a space station in low Earth orbit (LEO). Assuming China sustains an interest and financial wherewithal for such a task, the money is on a manned Chinese lunar expedition no sooner than 2025. But, when they do travel to the Moon, chances are they won't be following in American footsteps. They will be thinking bigger and better—a long-term stay and the first Moon colony. Perhaps the bigger threat may not be competition between China and the US, but between China and emerging commercial space ventures, like … Bigelow Aerospace—we'll discuss this more in Chapter 8.

Space stations

Before Bigelow realizes his lunar ambitions, there is the business of launching those inflat-ables. Bigelow's business plan calls for three full-size space stations (Figure 1.3) in orbit by 2017, and more than two dozen launches a year to service them. The company is target-ing two types of customer: sovereign clients who don't have their own space agency but who want a place to work in space, and companies involved in biotech and other industries who would sign leases to perform research. The business plan has the potential to revolu-tionize access to space. With an orbiting habitat for lease, all of a sudden, smaller countries with no manned spaceflight experience such as Norway or the United Arab Emirates will be able to take their first steps into space. And do it affordably and quickly. Bigelow's commercial space stations will also open up other opportunities. Nations such as Canada, Japan, Brazil, the UK, the Netherlands, and Sweden could secure the future of their human spaceflight programs and dramatically increase the size of their astronaut corps and increase the number of flights their astronauts fly. This move would be (very) popular among the astronauts who are lucky if they get two missions in a career nowadays. For example, Jeremy Hansen and David St. Jacques, Canada's newest astronauts, were selected in 2009 and the earliest that one of them will fly may be 2018. Then there are the smaller countries with no human spaceflight experience such as Singapore, which could take their first bold steps into space in a rapid (no lengthy delays dealing with government bureau-cracy) and affordable (see sidebar) fashion. The benefits to these nations are many. First, developing an astronaut corps and conducting operations aboard a space station can

1.3 Bigelow's inflatable habitat mock-ups. Courtesy: Bill Ingalls/NASA

dramatically transform a nation's image. Second, the creation of jobs and lucrative economic opportunities via microgravity research, development, and manufacturing can inspire a new cadre of scientists and engineers. Third, these commercial stations will present unique opportunities for corporations to gain advantages over their competition.

Astronaut Costs

US$26.25 million per seat on board SpaceX's Dragon, or US$36.75 million on Boeing's CST-100/Atlas rocket combination (compare this with the US$63 million cost of a Soyuz ride).

Cost of stay: lease block costs of US$25 million for exclusive use of 110 m^3 for up to two months.

Those individuals visiting the Bigelow station will enjoy 10–60 days in orbit. During this time, astronauts will be granted access to shared research facilities, which will include equipment such as a centrifuge, glove-box, microscope, furnace, and freezer. Unlike the ISS, where astronauts dedicate most of their time to house-keeping activities, astronauts on board Bigelow's station will be able to focus exclusively on their own activities, ensuring that nations and companies gain full value for their investment. For clients wishing to enjoy exclusive access to their own on-orbit facility, Bigelow customers can lease a third of one of its BA-330 habitats for 60 days, for only US$25 million. Another option for prospective clients interested in enhanced visibility are the naming rights offered by Bigelow Aerospace. For US$25 million, a corporation or nation can purchase the naming rights to the entire station for a year.

Tourism

Bigelow Aerospace is not in the business of running space hotels, despite what you have probably read in several popular science magazines. In fact, space tourism, together with military contracts, barely feature in the business plan. The reason the military isn't a concern is because Bigelow wants the stations to be used for peaceful purposes only. Basically, he wants his company to be a kind of orbital Hudson Bay Company, selling goods and services in the NewSpace economy.

Challenges

It all sounds promising, but private space stations have been attempted before. Joe Allen, a former astronaut who flew two Shuttle missions before leaving NASA in 1985 for the private sector, spent years working on a plan to build and operate a small space station. For a cost of US$1 billion to US$2 billion, the Industrial Space Facility (ISF) would have served as a commercial space station leased to NASA. Although the ISF was ultimately torpedoed by NASA's sloth-like decision-making and skeptical reviews, it gave Allen an understanding of the challenges of running a commercial space habitat. While the technical challenges are fairly routine, the business problems are anything but, especially with the "build and they will come" approach Bigelow has taken. The hurdle many space policy observers cite is these inflatables represent a major new technology, and history shows that these don't get developed fully without a subsidy or incentive: the railroads had free land they could commercialize, and the 1925 Kelly Act launched the airline industry by guaranteeing airmail routes to commercial carriers. The question is, without NASA or some other deep-pocket agency as a major anchor tenant, can Bigelow's private space business succeed? The naysayers don't deter Bigelow, who is determined to go it alone, without NASA if necessary. He is fond of comparing his space venture to real estate deals, saying that, in principle, his space station is little different from a US$50 million office building. Maybe. But an office building isn't surrounded by a vacuum, or bombarded by radiation and meteorites whizzing around at seven kilometers per second. NASA spends millions on technology to keep its astronauts safe. Why does Bigelow think he can do it for less?

Perhaps the key is the viability and versatility of the technology, although the idea of an inflatable space habitat isn't new: in the 1960s, Goodyear Aircraft drew up designs for a doughnut-shaped expandable orbital station (Figure 1.4) and, at one point, NASA

1.4 Goodyear Aircraft Corporation's space station concept. This design made it out of the concept phase and into production, though no models were flown. This station was designed to be launched into orbit and then inflated. In theory, the station could have been big enough for one or two crewmembers. Courtesy: NASA

considered using inflatable passageways to connect space modules. Today, Bigelow's inflatable habitats, each slightly larger than a shipping container and shaped like a giant watermelon, lie on the floor of a spaceship factory in North Las Vegas. Their bullet-proof skins, inflated by compressed, breathable air released from tanks in a steel core, are composed of numerous layers of flexible fabrics. The first is an air bladder that maintains the atmosphere. The bladder is surrounded by interwoven straps that hold it in the proper shape and ensure it doesn't burst. The outermost skin is five sheets of protective shielding, made of heavy-duty Kevlar and Vectran-like materials, which keep the habitat from being punctured by micrometeorites. The curving walls are set with portholes, and the space is divided into three floors by interlaced fabric bands. The attraction to potential clients is that the modules can be configured to whatever the user wants: crew quarters, experiment areas, exercise area, payload racks … you name it. It's precisely because of the versatility of these inflatable habitats that Bigelow may be able to make this business work and why he is winning over a constituency in the stodgy world of established aerospace corporations and government agencies—NASA included. His hobbies may raise eyebrows, but his successes have caught their attention. In the world of manned spaceflight, stranger things have happened.

Beyond inflatable habitats in orbit are lunar mining missions. The Moon contains large concentrations of helium-3, a nonradioactive isotope that is rare on Earth but could be a valuable fuel for nuclear fusion. Until very recently, acquiring those lunar resources was a difficult and expensive proposition, but Bigelow reckons the balance of risk and reward is changing with low-cost private rockets and his inflatable modules, which could be modified for surface use. Establishing a lunar base would also place Bigelow in competition with the Chinese, who have made no bones about why they are going to the Moon.

CAN A REAL ESTATE GUY ACTUALLY DO THIS?

In January 2013, Bigelow Aerospace and NASA announced they had signed a US$17.8 million fixed-price contract under which one of Bigelow's expandable modules would be attached to the ISS in May 2015. The module will be launched on the eighth Commercial Resupply Services (CRS-8) flight NASA already has under contract with SpaceX. The BEAM will berth to the ISS Tranquility node and operate in a closed hatch mode where ISS crewmembers will enter to check on experiments from time to time. Two years after docking, BEAM will be jettisoned, although, if things are going well, the mission might be extended. Compared to the hundreds of millions of dollars the company has invested,[3] the US$17.8 million NASA contract is chicken feed but, for Bigelow, the value of the agreement is working with NASA and starting a long-term relationship with the agency. Once in place, the module will serve as a technology demonstrator to determine how it withstands the radiation, debris, and thermal environment. Assuming the technology is robust enough to withstand the space environment, Bigelow plans launching two BA-330 models by the end of 2016 that will be available for lease. Will Bigelow's business model

---

[3] At the time of the NASA announcement, Bigelow Aerospace had invested about US$250 million in BEAM and expects to invest a similar amount by the end of 2016.

succeed? Perhaps. The market for privately funded space companies is real. Don't forget, since NASA retired the Shuttle, the agency has had to rely on foreign governments and private companies to get into space, including the transport of cargo and crew to the ISS. The US space business had estimated revenues of US$40.9 billion in 2010, according to the Aerospace Industries Association, up 18% from revenue of US$33.6 billion in 2005. By the end of the 2010s, revenues will be more than double what they were in 2010, according to aerospace consulting firm Ascend. The black-and-white truth of the matter in 2014 is that space is no longer just in the realm of government anymore. Nowadays, much of NASA's budget goes to commercial partnerships, signaling an era in which the relationship between the public agency and private industry has changed from: "Here's what I want, go build it" to "Here's what I want. Now let the private sector figure out the most cost-effective way". And the evidence that private industry can do the job as well or better than government abounds. Take SpaceX, for example. A NASA study estimated SpaceX spent US$390 million developing its Falcon 9 rocket and launch vehicle but, if NASA had done the same work, it would have cost between US$1.7 billion and US$4 billion. Incidentally, SpaceX may be ferrying Bigelow customers in the very near future.

2

Expandable Module Technologies

© Springer International Publishing Switzerland 2015
E. Seedhouse, *Bigelow Aerospace: Colonizing Space One Module at a Time*,
Springer Praxis Books, DOI 10.1007/978-3-319-05197-0_2

"Within the next 10 to 15 years, the Earth will have a new companion in the skies. An artificial moon, from which a trip to the moon itself will be just a step, carried into space, piece by piece, by rocket ships."

Wernher von Braun

INFLATABLE SPACE STATIONS

The idea of inflatable space stations may sound revolutionary, but this technology has a history stretching back even before the birth of NASA. The von Braun reference dates back to the great man's 1945 study for an American manned space station. The toroidal station (Figure 2.1), which spun to provide artificial gravity, became familiar to the American public over the next six years, and the design, which was elaborated at the First Symposium on Space Flight on October 12th, 1951, at New York's Hayden Planetarium, was popularized in *Colliers* magazine, and illustrated by famed space artist, Chesley Bonestell. The 1946 version used 20 cylindrical sections, each about three meters in diameter and eight meters long, to make up the toroid. The station spanned about 50 meters in diameter and guy wires connected and positioned the toroid to the central power module, which was fitted with a solar collector dish, designed to run an electrical generator. The 1952 version, which was enlarged to 75 meters in diameter and housed 80 crew, was

2.1 Von Braun's inflatable space station. Courtesy: NASA

improved by making the station's toroid out of smooth donut-shaped, inflatable sections made of reinforced rubber.

Inflatable technology was just one of many new technologies featured on the station.[1] The orbiting outpost would rotate to produce artificial gravity at the crew levels, which would feature two crew-height living and working areas, while the outer level would be dedicated to utilities. Space taxis would move from docking ports at the center of the station to arriving shuttles, and to conduct assembly operations of Moon-bound spacecraft near the station. Protecting the station's inflated torus would be a metal meteorite shield. The station would be in a 1,730-kilometer circular two-hour pseudo-Sun-synchronous orbit, meaning it would always be in sunshine as Earth revolved below it. Sporting a total volume of 18,400 cubic meters, the station would require 24 metric tonnes of a nitrogen/oxygen air mixture for pressurization, although use of a helium/oxygen atmospheric mixture would reduce the total mass of atmosphere aboard the station to 16 metric tonnes while eliminating the risk of bends in case of depressurization.

While von Braun's prediction that the station would become a reality in a few decades was not realized, the prospects of a space station did not pass unnoticed at NACA (National Advisory Committee for Aeronautics) Langley, where researchers speculated about the technology needed to develop an orbiting outpost such as the one von Braun had proposed. Then, in 1958, with the ink still drying on the Space Act that created NASA, interest in a space station ratcheted up considerably, and preliminary working groups concerned with space station concepts came alive within the newly founded agency and the aerospace industry. One of the working groups was NASA's inter-center Goett Committee, which met for the first time on May 25th–26th, 1959, to propose ideas for the next manned spaceflight objective after Project Mercury. One of the most enthusiastic members was Langley representative Larry Loftin, whose Project AMIS (Advanced Man In Space) presentation recommended "NASA undertake research directed towards the following type of system: a permanent space station with a 'transport satellite' capable of rendezvous with the space station". In his presentation, Loftin explained how the space station could serve as a medical laboratory to study the effects of space on astronauts, how researchers could study the effects of the space environment on materials, and how NASA could use the station to develop new stabilization, orientation, and navigational techniques. The minutes of the Goett Committee do not record the immediate reaction to Loftin's presentation, but several members of the committee had a strong feeling that a manned space station should be the project after Mercury.

To that end, the Manned Space Laboratory Research Group was formed. It consisted of six subcommittees responsible for the study of various aspects of space station design and operations: (1) Design and Uses of the Space Station; (2) Stabilization and Orientation; (3) Life Support; (4) Rendezvous Analysis; (5) Rendezvous Vehicle; and (6) Power Plant. The Group's goal was to send an astronaut to the Moon and back using an orbiting station as a launch pad for the lunar landing mission. By the fall of 1959, work had progressed to the

---

[1]   Writer Arthur C. Clarke and director Stanley Kubrick borrowed the torus design for their classic 1968 movie *2001: A Space Odyssey*. In one of the most iconic scenes of the film, the space wheel turns majestically to the waltz of Johann Strauss's *The Blue Danube* while a Pan Am shuttle with passengers aboard approaches the station.

point where Loftin made a three-point statement of purpose: Langley would study the psychological and physiological reactions of man in space over extended periods of time, provide a means for studying materials, structures, and control and orientation systems, and study means of communication, orbit control, and rendezvous.

SATELLOONS

At about the same time as the Manned Space Laboratory Research Group was formed, work was already underway developing inflatable structures for space—satelloons.[2] As the civilian agency responsible for US space activities, NASA has had a program of technology development for satellite communications ever since the agency was established in 1958. One of the agency's first projects—Echo—was to develop a passive communications satellite that reflected radio waves back to the ground. The Echo project began in 1956 as a NACA experiment to test the effects on large lightweight structures in orbit. Then, in 1958, when NASA was created and NACA dissolved, Echo (Figure 2.2) became a NASA project.

Built by the G.T. Schjeldahl Company (Grumman built the dispenser), the Echo satellite was a 31-meter-diameter aluminized-polyester balloon that inflated after orbital insertion. The balloon's development was enabled by a 12.7-micrometer-thick metalized biaxially oriented polyethylene terephthalate (PET) film material known as Mylar® (see sidebar). During ground inflation tests, up to 18,000 kilograms of air was needed to fill the 68-kilogram balloon but, once in orbit, several kilograms of gas were all that was required to fill the sphere. To ensure the balloon remained inflated in the event of micrometeorite strikes, a make-up gas system using evaporating liquid was integrated inside the satellite. It was also fitted with 107.9-megahertz beacon transmitters for telemetry purposes, powered by five nickel-cadmium batteries charged by 70 solar cells mounted on the balloon. Following the failure of the Delta rocket carrying Echo 1 on May 13th, 1960, Echo 1A[3] was placed into a 1,519–1,687-kilometer orbit by another Delta rocket on August 12th, 1960, and two-way voice links were set up between Bell Telephone Laboratories in Holmdel, New Jersey, and NASA's Jet Propulsion Laboratory (JPL) facility at Goldstone, California. The program was successful, since Echo demonstrated satellite tracking and ground station technology that was later applied to active satellite systems. Buoyed by the success of the world's first inflatable satellite, it wasn't surprising that Echo 2 was built. Managed by NASA's Goddard Space Flight Center in Beltsville, Maryland, Echo 2 was launched on January 25th, 1964. Fitted with an upgraded inflation system, which improved the balloon's smoothness, Echo 2's investigations were concerned more with the dynamics of large spacecraft.

---

[2]   The satellite was nicknamed a "satelloon" by those involved in the project, as a portmanteau of satellite–balloon.

[3]   It finally re-entered Earth's atmosphere and burned up on May 24th, 1968.

2.2 Echo satelloon. Courtesy: NASA

Echo 2 was a 41.1-meter-diameter, metalized PET film balloon, which was the last satelloon launched by Project Echo. Launched on January 25th, 1964, on a Thor Agena rocket, it used an enhanced inflation system to improve the satelloon's sphericity, and sported a beacon telemetry system that provided a tracking signal, monitored spacecraft skin temperature (between −120°C and +16 °C), and measured the spacecraft's internal pressure. The system consisted of two beacon assemblies powered by solar cell panels and had a minimum power output of 45 milliwatts at 136.02 megahertz and 136.17 megahertz. In addition to the passive communications experiments, Echo 2 was used to assess the dynamics of large spacecraft and for global geometric geodesy. After the satelloon re-entered Earth's atmosphere and burned up on June 7th, 1969, NASA abandoned passive communications systems in favor of active satellites.

Mylar

Boasting a diameter close to the height of a 10-storey building, the Echo satelloons have been described as perhaps the most beautiful objects ever to be put into space. The challenge NASA engineers faced was how to place such a large structure into the tiny Thor-Delta fairing—a challenge that was to later inspire the TransHab invention. The solution was to use an inflatable system, which led to the satellite being made out of Mylar®. Mylar® polyester film was invented in the early 1950s and its use in the Echo project was just the first of many firsts in the space industry, with variants being used in space blankets and as linings in spacesuits.

INFLATABLE SPACE STATIONS: THE SATELLOON LEGACY

As the Langley researchers got to work examining the feasibility of various space station configurations in 1960, they soon agreed that the most promising design was a self-deploying inflatable. Thanks to their satelloon experience, Langley engineers already knew first hand that a folded station packed snugly inside a rocket would be protected during the rough ride through the atmosphere. That's not to say the Langley space station office didn't consider other concepts. They did. Among the non-inflatable concepts considered were designs for orbiting cylinders, and for a cylinder attached to a terminal booster stage, but these were rejected as dynamically unstable because they tended to roll at the slightest disturbance. A version of Lockheed's elongated modular concept was turned down because it required the launch of several boosters to place all the elements into orbit, and proposals for Ferris wheels in space were rejected because of Coriolis effects (see sidebar). While the Langley space station team had sound technical reasons for doubting the feasibility of these proposals, it perhaps wasn't surprising they favored the inflatable option because the technology was developed at Langley! The concept also happened to make good engineering sense because the inflatable option meant a light payload and, with hundreds of kilograms of propellant required to put just one kilogram of payload into orbit, any plan that lightened the payload was a winner.

Coriolis Effects: A Primer

A rotating space station will produce the feeling of gravity because the rotation drives any object inside the station towards the hull. This "pull", or centrifugal force, is a manifestation of objects inside the station trying to travel in a straight line due to inertia. From the perspective of astronauts rotating with the station, artificial gravity behaves similarly to normal gravity, but there are side effects, one of which is the Coriolis effect, which gives an apparent force that acts on objects that move relative to a rotating reference frame. This force acts at right angles to the motion and the rotation axis, and tends to curve the motion in the opposite sense to the station's spin. If an astronaut inside a rotating station moves towards or away from the axis of rotation, they will feel a force pushing them towards or away from the direction of spin. These forces act on the inner ear and can cause nausea and disorientation. Slower spin rates (less than two revolutions per minute) reduce the Coriolis force and its effects, but rates above seven revolutions per minute cause significant problems.

The first idea for an inflatable station was the Erectable Torus Manned Space Laboratory, developed by the Langley space station team led by Paul Hill and Emanuel Schnitzer with the help of Goodyear. Their idea was a flat inflatable unitized torus about seven meters in diameter. Since it was unitized, all its elements were part of a single structure that could be carried to orbit on one booster, which was a major selling point. All NASA needed to do was fold the station into a compact payload. The Langley space station team was so enthusiastic about its inflatable torus that they made a presentation on the design to a national meeting of the American Rocket Society. In the months following their presentation, Langley built and tested models of the Erectable Torus Manned Space Laboratory (Figure 2.3), including a full-scale research model constructed by Goodyear.

Development of the concept appeared promising, but the design had its drawbacks. For one thing, engineers worried that if the flexible material was not strong enough, astronauts moving around vigorously in the space station might somehow propel themselves so forcefully they would break through the fabric and shoot into space! There was also the more serious engineering concern related to the dynamics of the toroidal structure. When arriving astronauts moved equipment from the central hub to a working area at the outside periphery, it was suggested the station might become unstable, thereby upsetting its orbit. So, knowing astronauts working in the station would have no weight but would still have mass, the Langley engineers conducted studies to calculate the effect of astronauts moving about in the habitat. The results showed the mass distribution would be changed when crewmembers walked from one part of the station to another, producing a slight oscillation of the station. The next step was to investigate whether it was possible to alleviate the oscillation, so the Langley engineers built a three-meter-diameter elastically scaled model of the torus. By the time the model became operational in 1961, NASA had realized it had to either develop a more rigid inflatable or abandon the inflatable idea altogether.

2.3   Erectable Torus Manned Space Laboratory. Courtesy: NASA

    While still pursuing the inflatable torus concept, the Langley group also explored other ideas. In the summer of 1961, it entered into a six-month contract with North American Aviation for a feasibility study of an advanced modular space station concept, which also incorporated inflatable technology. While rigid in structure, this advanced station could still be automatically erected in space. The idea was to put together six rigid modules connected by inflatable spokes or passageways to a central non-rotating hub. The 22.8-meter-diameter structure would be assembled on the ground, packaged into a snug launch configuration, and launched into space. To ruggedize it against micrometeorites, the rigid sections of the rotating hexagon airlock doors could be sealed when any threat arose to the integrity of the interconnecting inflatable sections. The structure was designed to rotate, making it possible for astronauts to take advantage of artificial gravity, which space station designers of the day believed was an absolute must for any long-term stay. Incidentally, the 22.8-meter-diameter size was selected because it provided the minimal rotational radius needed to generate the 1 G required for the station's living areas.
    As the Langley engineers continued to investigate the potential of a rotating hexagon, they became increasingly confident they were on the right track. The only problem was finding a launch vehicle capable of lofting the 77,500-kilogram structure into orbit.

The solution was von Braun's Saturn, so a team of Langley researchers tried to figure out how to mate their space station to the Saturn's top stage. After working with a number of dynamic scale models, they refined a system of mechanical hinges enabling the six interconnected modules of the hexagon to fold into one compact mass. Tests confirmed the arrangement could be carried aloft in one piece and, once on orbit, actuators located at the joints between the modules would deploy the folded structure. The cost for the space station project was US$100 million. At the time, this was too much for NASA, which only had sufficient funding to finish Mercury and US$29 million for Apollo. Also, NASA wasn't even sure it needed a space station, because Apollo entailed only a circum-lunar mission, with the possibility of building a space station as a by-product of the Earth-orbit phase. Such uncertainty is par for the course in the aerospace industry, but it put Langley in a difficult situation. Since some sort of space station was still possible in the Apollo era, the basic technology had to be ready, so Langley continued their research. On May 19th, 1961, six days before President Kennedy's lunar landing speech, Loftin updated the US House Committee on Science and Astronautics on Langley's manned space station work. He passed around a series of pictures showing Langley's concepts for the inflatable torus and the rotating hexagon, before summarizing Langley's assessment of the status of the space station. The politicians were somewhat flummoxed, many of them not understanding what a manned space station was all about or how it might be used.

Six days after Loftin's appearance, President Kennedy stunned the world—including NASA—with his lunar landing speech. For 14 months following Kennedy's speech, NASA debated various mission modes. Many in NASA were certain the mission architecture would involve Earth-orbit rendezvous, which would require the lunar spacecraft to be assembled from components put into orbit by two or more Saturn rockets. This plan would therefore involve the development of orbital capabilities that might translate into a space station. With this in mind, the Langley team continued to explore the problems facing the design and operation of a space station. One continuing issue was how to protect astronauts from micrometeorite strikes, because big hits, especially those striking the inflatable torus, could prove disastrous. In an attempt to solve the problem, structure experts at Langley and Ames searched for a wall structure that offered the greatest protection for the least weight. They settled on a sandwiched structure with an inner and outer wall—a precursor to the layered structure that was later used in TransHab. Developed by North American, the outer wall was a meteorite shield comprising aluminum, backed by a polyurethane plastic filler that overlaid a bonded aluminum honeycomb sandwich. The wall seemed rugged enough to do the job, but no one really knew because there was no way to simulate micrometeorite strikes in any ground facility. For the inner wall, Langley's engineers decided nylon-neoprene, Dacron-silicone, saran, Mylar, polypropylene, Teflon, and other flexible and heat-absorbing materials could do the job. What made these materials attractive was their ability to withstand a hard vacuum, electromagnetic and particle radiation, and large temperature changes. At a symposium in July 1962, the Langley team presented summary progress reports on their space station research, concluding that the rotating hexagon was superior to the inflatable torus. The inflatable concept was still a possibility. But not for long.

BIRTH OF TRANSHAB

In 1963, an inflatable extension was proposed for the Apollo program but, by that time, technology based on hard aluminum shells had became prevailing, although suggestions of how to use inflatable technology were still discussed. For example, when NASA embarked on the development of its second-generation manned spacecraft called Gemini, engineers considered an inflatable delta wing as an alternative to the primary landing method of splashing into the ocean with parachutes. Another suggestion was an inflatable paraglider (Figure 2.4), which promised a controlled landing of a two-seat capsule on land, but the pressures of the space race left precious little time to resolve the technical challenges of the new system.

But NASA didn't do away with inflatable technology completely: inflatable air bags (Figure 2.5) were used on the Command Module of the Apollo spacecraft to ensure a vertical position following water landing. And, on the subject of the space race, the Americans weren't the only ones who recognized the advantages of inflatable structures in space: in 1965, Alexi Leonov conducted the first spacewalk from a cylindrical inflatable airlock fitted on board the Voshkod-2 spacecraft. The event almost ended in tragedy when a miscalculation in the pressurized size of Alexi Leonov's spacesuit caused him great difficulty when re-entering through the airlock's small hatch.

After developing its space station, Goodyear continued its research into the application of inflatable structures by proposing a lunar shelter (Figure 2.6) designed to support a crew of two for periods of 8–30 days at a pressure of 0.35 kilograms per cm^2. The shelter's outer and inner layer materials were polyaramid nylon fabric bonded by polyester adhesive to provide micrometeorite protection. A middle layer was a closed-cell vinyl foam for radiation protection and thermal insulation. The module and airlock, whose volume was 14.5 cubic meters, was constructed of a three-layer laminate consisting of nylon outer cover, closed-cell vinyl foam, and inner nylon cloth bonded by polyester adhesive layers.

Goodyear then developed a larger space module prototype for a proposed 36-meter-long lunar base habitat in 1968. The outer surface was covered with a nylon film-fabric laminate covered with a thermal control coating, and the inner layer was a gas bladder made from PET dipped in a polyester resin bath, and sealed by a polyvinyl chloride (PVC) foam. The middle layer was flexible polyurethane foam. Designed to operate at a pressure of 0.35 kilograms per cm^2, the entire structure weighed just 735 kilograms!

Next on the Goodyear drawing board was a two-meter-long inflatable airlock designed to be mounted on a Skylab-type vehicle. Developed through a joint NASA/Department of Defense venture in 1967, the structural layer used a thin filament-wound wire for tensile strength while flexible polyurethane foam provided a micrometeorite barrier, and a fabric-film laminate afforded thermal control. The compact unit weighed just 83 kilograms and fit into a snug 1.2-meter-diameter, 76-centimeter-tall cylinder. In the course of its research into inflatable applications, Goodyear also qualified a flexible fabric consisting of Nomex unidirectional cloth coated with Viton B050 elastomer. The combination offered potential applications for habitats because the Nomex/Viton structural layers could be laminated together for strength, and a flexible cable could serve as a bead to ensure structural integrity during deployment and when a structure was inflated. Another concept developed at the time was the rigidization of structures to ensure the volume of the habitat was retained

2.4   Inflatable paraglider concept. Courtesy: NASA/Smithsonian Air and Space Museum

2.5   Airbags used on the Apollo capsule. Courtesy: NASA

2.6   Lunar shelter concept. Courtesy: NASA

after inflation gases had been used. It was suggested rigidization might be accomplished by incorporating a flexible mesh core material impregnated with a gelatin resin between membranes of a sealed structure which expanded to harden the core when the wall cavity was vented to space vacuum during structure deployment—some of this technology laid the groundwork for NASA's Surface Endoskeletal Inflatable Modules, which are discussed in Chapter 8.

In addition to Goodyear, ILC Dover, developer of spacesuits, also became a leader developing advanced technology inflatable systems, including a hyperbaric chamber that had similarities to space habitats. The 0.8-meter-diameter, 2.1-meter-long structure included a bladder layer to retain pressure, and a restraint layer to support structural loads. The bladder comprised a urethane-coated polyester, and the restraint was a series of polyester webbings stitched to a polyester fabric substrate.

Meanwhile, Apollo came and went and very little happened in the world of inflatable structures. For many years, the Langley engineers hoped the idea of inflatable modules would catch on, but the idea was cast aside, not because people doubted the technology, but because nobody championed the cause. The result was that, by the mid-1970s, the development of inflatables had come to a halt at NASA. The concept was revived briefly in the late 1980s when the Bush Administration talked about returning to the Moon and Mars, issuing National Security Presidential Directive 6, authorizing the Space Exploration Initiative (SEI). At the time of the SEI, the fundamental advantages of inflatable structures still held true (a large amount of volume packed into a slim rocket fairing equals savings on mass and cost), which is why the technology was once again considered by NASA engineers. Faced with the same problem as their predecessors in the 1960s, NASA needed to get a significant amount of volume into space and had only a limited amount of rocket fairing room with which to do so. The solution? An inflatable crew habitat. In 1989, one of the planetary exploration concepts NASA proposed was the Inflatable Habitat Concept for a Lunar Base (Figure 2.7).

In the same year, the Lawrence Livermore National Laboratory studied the feasibility of inflatable modules to be used in a future space station (echoing the concept von Braun suggested almost four decades earlier) or possibly integrated in the space station Freedom, which was moving into hardware fabrication phase. The cylindrical and toroidal shapes investigated in the study were five meters in diameter and approximately 17 meters long. The dimensions of the deployed modules were not much bigger than the Shuttle's cargo bay, but they offered big savings on weight and take-off volume. A prototype inflatable sphere was developed in 1989 as part of the SEI initiative, but it kept tearing along the seams. Perhaps as a result of the problems encountered with the sphere, concepts for inflatable Moon lodges and construction shacks never made it much further than the proposal stage, although the inflatable cause was far from dead.

In 1996, NASA's Johnson Space Center (JSC) began to study a Moon mission that envisioned the use of an inflatable habitat to support checkout activities before a permanent habitat was established. ILC Dover was contracted to study various configurations and sub-assemblies including bladder, restraint layer, and thermal and micrometeoroid layers. The system was designed to sit atop a landing craft and expand on the surface. A number of concepts were assessed for the lunar surface module's construction. These included a rugged bladder design that was a dual-walled self-sealing silicon-coated

2.7   Inflatable Moon base concept. Courtesy: NASA

Vectran fabric with film laminates that afforded simplicity and cold temperature deployment properties. Several restraint layer concepts were also investigated, including coated single-layer fabrics, layers with circumferential and axial webbing over coated fabric, and structures with circumferential toroidal webbing over an internal axial layer. The selected wall system comprised a restraint layer that applied an outer Kevlar layer overlaying a structural denier plain weave.

Lowell Wood, a physicist at Lawrence Livermore National Laboratory, was one of the more vociferous inflatable proponents. He had been pushing inflatables as a way for this lab to take advantage of the SEI. Sadly, like so many NASA initiatives, SEI was plagued by a lack of budgetary and political support, and eventually went nowhere. But the idea of an inflatable crew habitat lived on and was revived as a crew quarters for the International Space Station (ISS). Despite the change in purpose, NASA again began focusing its attention on an inflatable habitat, conceived in early 1997 by a JSC engineering team under the direction of William Schneider, who had worked on micrometeorite protection for the Shuttle. Until Schneider's arrival, micrometeorite protection had been the inflatable's Achilles heel, but Schneider's team solved that by devising a Nextel–foam combination, which is why Schneider generally gets credit for being the Father of TransHab. The advantages of TransHab were obvious (Table 2.1). Like all inflatable systems, TransHab offered

Table 2.1. TransHab by the numbers.

Weight at launch	13.2 tonnes
Length at launch	11 meters
Diameter at launch	4.3 meters
Diameter after inflation	8.2 meters
Inflated volume	339.8 cubic meters

a huge amount of on-orbit volume while taking up a small amount of rocket fairing space relative to a traditional metallic structure. Additionally, the habitat provided enhanced protection from radiation compared with traditional metallic habitats. The reason is this. When exposed to cosmic rays or solar flares, metallic habitats may suffer from damaging secondary radiation wherein the metal creates a scattering effect. In contrast, due to nonmetallic material being used as the primary envelope, inflatables can significantly reduce this dangerous phenomenon. But could the inflatable habitat be adapted for the ISS? The team was given a year and US$2 million to find out. They got to work quickly, and submerged a mock-up in the Neutral Buoyancy Laboratory where it was overfilled with air and subjected to four times the maximum operating pressure of the station. It passed with flying colors. Incidentally, the space station's modules are designed to hold two times normal pressure.

Next was a vacuum test. The team assembled a test article in the Apollo-era vacuum chamber in Building 32 at Johnson (Figure 2.8). The test article was a slightly smaller version than the flight mode, which would be 8.2 meters in diameter and 11 meters high (if a human centrifuge was approved, the module would grow by another meter in height). The flight model would weigh about 11,800 kilograms empty and 15,875 kilograms outfitted with water. Compare these dimensions with Boeing's US Laboratory Module, which would have been 4.5 meters in diameter, 8.5 meters long, and weighed 14,500 kilograms. The team conducted a vacuum test that confirmed the module could fit into the Shuttle bay and unfold and inflate afterwards. The module was inflated to space station pressure of 1.04 kilograms per centimeter and, except for some restraining cords being hung up in the eyelets, it performed perfectly.

At the time of the tests, the price of a TransHab was US$200 million, compared to the US$300 million it had cost Boeing to build the Unity node. One of the reasons NASA didn't commit to the inflatable option was Boeing. As a prime contractor, the aerospace behemoth wasn't going to give up easily, which, for the astronauts and cosmonauts, was a shame. The sleeping quarters in TransHab would have been 25% bigger than those in the Boeing habitation module and the quarters could have been used as a storm shelter thanks to the surrounding water. Apart from the cost savings (a poor man's way of getting more volume), TransHab also represented pioneering technology and a bridge to exploration down the road—a philosophy that sat particularly well with the astronauts. Sadly, despite showing great technical promise, when it was revealed that the ISS program was US$4.8 billion over budget, the TransHab/crew habitat program was canceled by Congressional bean-counters in 2000.

2.8   Vacuum chamber with TransHab test article. Courtesy: NASA

THE RUSSIAN AND EUROPEAN INFLATABLE EXPERIENCE

Despite Leonov's harrowing experience with the inflatable airlock, the Soviets pressed on with developing the technology. In 1966, after many failures, the Soviet Luna-9 unmanned probe soft landed on the Moon thanks to the use of inflatable air bags that softened the impact onto the lunar surface. This method played a lasting role in planetary exploration in the USSR and in the US, and to date remains the most significant application of inflatable technology for space exploration.

In 1984, Soviet scientists placed instrument-carrying balloons on a pair of Vega space-craft heading to Venus. Following their flyby of the planet, Vega probes dropped re-entry capsules, which in turn released a pair of traditional landers and inflatable balloons to float in the atmosphere. Then, in 1996, the landing system first proven during the Luna-9 mission was resurrected in the post-Soviet Mars-96 mission. A pair of landers carried on board the main spacecraft were to land on the surface of Mars. Shortly before reaching the surface, a pair of airbags would inflate around each lander to soften the impact. The spacecraft would bounce and continue to bounce until they finally came to rest. Lines holding the two bags would be cut and the lander would free fall to the ground. In addition to landers, the Mars-96 spacecraft also carried a pair of penetrators—needle-shaped vehicles designed to strike the surface of the planet and penetrate four to six meters into the soil. After braking in the Martian atmosphere with the help of an inflatable heat shield, the penetrators were expected to strike the surface at a speed of about 60–80 meters per second. Unfortunately, Mars-96 was stranded in Earth orbit and neither the inflatable bags nor the inflatable heat shields on its penetrators had a chance to prove themselves.

In the same launch window for Mars-96, NASA launched the Mars Pathfinder mission, which utilized inflatable bags to ensure a soft landing. The spacecraft completed a flawless trip to Mars and the inflatable airbag system successfully delivered a lander and a small rover on the surface of the Red Planet in July 1997. Seven years later, NASA used the inflatable cushioning system again, delivering a pair of Mars Exploration Rovers onto the surface of the Red Planet.

Back in Russia, many space projects faced a budget crunch in the wake of economic problems of the post-Soviet period, but the European Space Agency (ESA) reckoned the Russian inflatable heat shield system from the Mars-96 project had potential, possibly as an affordable method of returning cargo from the ISS. To that end, ESA co-funded the Inflatable Re-entry and Descent Technology (IRDT; Figure 2.9) project together with the European Commission and Daimler Chrysler Aerospace, DASA, and, in February 2008, a pair of IRDT devices were launched on a Soyuz rocket. The smaller IRDT device was to return an experimental package from orbit, while the larger IRDT device was attached to the Fregat upper stage to protect the Fregat during its re-entry. After the flight, both IRDT devices successfully inflated and re-entered, but a radio beacon failure on both payloads coupled with bad weather at the landing site hampered search efforts and only the small device was recovered.

More recently, the company that built Leonov's spacecraft has jump-started work on multilayered inflatable structures. In its annual report for 2012, RKK Energia said the new project might pave the way for a new generation of space station modules, interplanetary spacecraft, and interplanetary bases. According to RKK Energia, inflatable modules will not only provide three times more volume and one and a half times more surface area per unit of mass than metal structures, but also promise lighter micrometeorite and radiation shielding. On the Russian ISS segment, an inflatable module could increase comfort for the crew and also provide increased volume for science experiments, including a centrifuge to create artificial gravity. RKK Energia began development of the inflatable module in 2011 using its own funding, in the hope of getting a future contract for such a structure from the Russian space agency, Roskosmos. During 2012, RKK Energia evaluated two sizes of the module, which could be launched on the Soyuz-2-1b rocket or on the Proton-M and Angara-A5 rockets. In the course of the project, RKK Energia procured domestically produced materials suitable for the inflatable module and developed a structural design and composition of the module's flexible skin. Samples of the materials were tested and a

2.9   Inflatable Re-entry Descent Technology. Courtesy: NASA

scale model was used to assess the module's skin. The troublesome issue of micrometeorite protection was dealt with by a research center at Roskosmos (TsNIIMash research institute), which developed an undisclosed protective material that was tested alongside traditional AMG-6 aluminum alloy used in the aerospace industry. According to the company, the new materials provided 95% of the required level of protection. During 2014, RKK Energia and its contractors planned to build a one-third-scale prototype of the module for ground tests.

TRANSHAB: A PRIMER

Going back to the late 1990s, practically everyone thought TransHab was cool, and potentially very useful, but it didn't fit into NASA's plans for the space station and was abandoned. We'll get into the technical aspects of this invention in the next chapter but, before we do, it's useful to have an understanding of the basic structure, so what follows is an overview of this inflatable technology. At its most basic, TransHab is a unique hybrid structure combining the packaging and mass efficiencies of an inflatable structure with the advantages of a load-bearing hard structure. TransHab's inflatable shell comprises multiple layers of blanket insulation, protection from orbital and meteorite debris, an optimized restraint layer, and a redundant bladder with a protective layer (Figure 2.10). With almost two dozen layers, the structure's inflatable shell is as unique and tough a design as they get: the outer layers are layered to break up particles of space debris and micrometeorites that may hit the shell at speeds of several kilometers per second, while the shell provides insulation from temperatures in space that can range from +121°C in the Sun to −128°C in the shade.

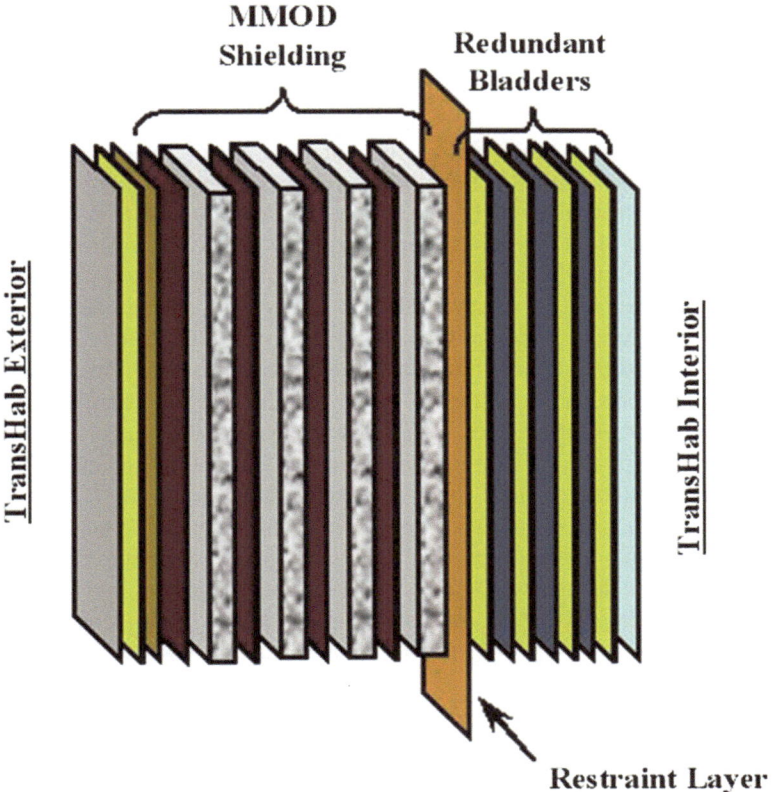

2.10 TransHab layers. Courtesy: NASA

The inflatable shell, which is TransHab's primary structure, is composed of four functional layers: the internal scuff barrier and pressure bladder, the structural restraint layer, the micrometeoroid/orbital-debris shield, and the external thermal protection blanket. The shell is folded around the core at launch and deployed once on orbit. Its function is to provide the crew with living space, and provide debris protection and thermal insulation. The key to the rugged protection is successive layers of Nextel, a material used as insulation under the hoods of many cars, spaced between several-centimeter-thick layers of open cell foam, similar to that used in chair cushions. Particles hitting at hypersonic speeds expend energy and disintegrate on successive Nextel layers, spaced by the open cell foam. For extra protection, the shell includes a thin layer of Kevlar. The layering has been tested extensively and repeatedly,[4] including having projectiles fired at the fabric sandwich at speeds of seven kilometers per second.

Underneath the outer shell is the restraint layer, woven from 2.54-centimeter-wide Kevlar straps. This layer is just as tough as the outer shell but it's a different type of toughness because the restraint layer's job is to contain pressure—four atmospheres of it. Under the restraint layer is an inner liner of Nomex that provides fire retardance and abrasion protection, and then there are three Combitherm (a material commonly used in the food packaging industry) bladders that form redundant air seals.

TransHab is a fabric construction, so it's hardly revolutionary, but the idea of launching such a structure into low Earth orbit (LEO) strikes many people as pretty radical—perhaps it's because the word "inflatable" is often used when describing the technology: NASA discouraged use of the word because it conjures up images of balloons. But let's focus on the fabric element. Many thousands of years ago, cavemen created temporary housing using animal skins stretched over bones while following herds of animals in their search for food. Over the centuries, other fabrics were developed to enhance similar structures and to fashion tents. More recent is the appearance of tensile fabric structures, which are now at the forefront of architecture with their dynamic shapes, sweeping boldness, and technological flexibility. If you watched the 2012 Olympics, chances are you saw several examples of what can be accomplished using tension fabric structures: the basketball arena was the largest temporary structure of the Games: the velodrome featured interior permeable fabric screening, and the press bar was an exterior canopy shelter. NASA understood the value of fabrics more than 50 years ago, just as it understood the value of the inflatable concept but, as is often the case when developing radical technology, the concept was mothballed not once (when the Moon program was halted), but twice (following the SEI). NASA didn't stop work developing fabric habitats; it just decided to rely mostly on aluminum because it was a known, safe, and reliable material. But, eventually, inflatable habitats made their way back to the table with the idea of attaching a module to the ISS. Over the years, NASA engineers continued to refine their ideas and concepts for inflatable structures until, one day in 1997, a Tiger Team was formed to design an interplanetary vehicle habitat for a crew of six to travel to and from Mars. The team, which comprised half a dozen engineers and a space architect, was led by Dr. William Schneider. The team identified one major challenge: how to deliver the habitat into space using

---

[4]   These tests were so dramatic that the cable-TV show *Scientific American Frontierp*, with host Alan Alda, included the test shots as part of a television series on Mars mission technology.

existing launch vehicles. Because of the considerable volume required per crewmember, for food, equipment, and consumables, the only viable solution was to use an inflatable structure. One member of Schneider's Tiger Team was Kriss Kennedy, a space architect at NASA's JSC. He had been working on the Mars Mission Studies and had already helped design various types of habitats as transit and surface habitats. When the team began designing the inflatable transit habitat, it was Kennedy who coined the name TransHab (short for *transit habitat*).

The original ISS TransHab—had it been approved—was to be carried in a compressed state on board a Shuttle and inflated once the Orbiter reached the outpost. TransHab would have provided a huge amount of space compared with traditional hard-shelled modules. The interior (Figure 2.11) would have featured three spacious levels—the lower level holding a dining area, the middle level a mechanical room and living quarters, and the top level an exercise room and shower. Water tanks would surround the middle level where the living quarters were and two portholes allowed views of Earth and the stars. With all these fancy options, you may be wondering why the technology didn't catch on, especially since NASA had been studying them for decades. One of the biggest hurdles was psychological because most people just didn't—and some still don't—believe it was possible for fabric to be as strong or stronger than metal. Even astronauts were among the early skeptics although, once they saw what TransHab had to offer in the way of living space, they became more supportive of the project. It was the concern over the use of the word

2.11 Deputy NASA Administrator Lori Garver next to the interior of a BA-330, a descendant of TransHab. Courtesy: Bill Ingalls/NASA

"inflatable" that perhaps caused the most concern, prompting Kriss Kennedy to pull out a balloon and pop it whenever he gave a public talk about the project. The popping was mainly to get the audience's attention, after which Kennedy would patiently explain that a balloon is a balloon and an inflatable structure is *not* a balloon. Horacio de la Fuente, the project's deputy manager for technical development, preferred a football analogy. He would explain that a football and TransHab have bladder systems to hold air. In fact, TransHab has three bladders covered by alternating layers of ceramic fabric, polyurethane foam, and Kevlar.

While talks by Kennedy and de la Fuente helped persuade skeptics TransHab was safer than it appeared, people really started to wake up to the integrity of the inflatable's space worthiness following the ballistic tests. The target of the tests was TransHab's outer layer, comprising mustard-colored Kevlar webbing (woven by hand to reduce the number of seams, thereby adding strength) and sheets of ceramic fabric (Nextel) separated by layers of foam. To test the module's ability to protect the crew from micrometeorite strikes, the TransHab team fired marble-sized aluminum balls at the shell at velocities of seven kilometers per second. Each time, the Nextel decimated the balls before they reached the air-containing bladders. Then the team turned their attention to the reinforced aluminum plates used in the ISS and performed the same test. The result wasn't pretty: the balls ripped chunks off the back of some of the plates, which were more than three centimeters thick, and left craters that left little to the imagination of an astronaut.

Many people were disappointed when development ceased on TransHab, but the project was just one of several trade studies performed to determine which habitat choice was best suited for ISS budgetary constraints. To that end, the TransHab project was intended to perform the design studies and engineering tests necessary to prove an inflatable module was viable for the ISS. Nothing more. And the project was extremely successful because it proved the viability of an inflatable option. So why was TransHab canceled? After all, in terms of long-term and operational costs and needs, there was no question the module would have saved the program tens of millions of dollars over the first decade, and would have given the outpost several modules that it could not otherwise have had. These included a safe-haven shelter for solar storms, on-orbit water-recycling capability, more than double its current total stowage volume, and the ability to test new exploration-class technologies like a human centrifuge, advanced medical facilities as well as the orbital-debris shielding for which TransHab has become famous. TransHab would also have furthered the progress of a manned Mars mission because NASA's Human Exploration and Development of Space (HEDS) roadmap defined these technologies as essential developments for any long-duration expeditions beyond Earth orbit.

Unfortunately, in early 2001, there was a US$4.8 billion shortfall in the ISS budget which had wide-ranging effects on NASA's space program, almost all of them damaging. One of the first things to be cut from the program was the habitation module and the Crew Return Vehicle which would be necessary to allow the crew to grow from three to seven. It was a shame, because the TransHab group had overcome major obstacles that had frustrated earlier attempts to build large inflatable modules. For example, the group developed the specialized weaving pattern that permits straps of woven Kevlar to withstand remarkable amounts of stress. For astronauts, the biggest attraction of living in a TransHab would have been the space. Lots of it. Take a look inside the current ISS (Figure 2.12) and what

2.12 The cluttered interior of the ISS. Courtesy: NASA

do you see? Modules packed with electronics racks, cables streaming everywhere, piles of manuals laying all over the place, and loud cooling fans running constantly. Equipment scattered all about, bulkheads covered with stuff attached to Velcro. "Cluttered" is perhaps the most apt word to describe living on board the ISS. Not exactly great working conditions! Now imagine living in that kind of environment 24 hours a day for several months and compare that living experience with the luxurious confines of a TransHab, which had a layout that would have encouraged a more orderly, productive, and pleasing environment to live and work in. It wasn't to be. Not in the late 1990s at any rate.

To be fair, part of the reason for the cluttered ISS is a legacy of the basic ISS architecture, which dated to the mid-1980s. Another problem is engineering. Ask any engineer which is more important—clutter and noise or dealing with a hard vacuum and 500-degree temperature fluctuations of space—and you realize where most of the money went. So, today, the ISS has done little to alter people's perception that working in space is little different from living in a submarine—dark, dank, and claustrophobic.

3

TransHab Up Close

In May 2015, NASA's BEAM (Bigelow Expandable Activity Module) mission will feature the robotic berthing of one of Bigelow Aerospace's BEAM modules at one of the interfaces of the International Space Station's (ISS) Tranquility node. As well as providing much-needed logistics and stowage support for the orbiting outpost, BEAM is expected to provide useful data on the performance of non-rigid space station modules. As discussed in the previous chapter, expandable space station modules are nothing new. But the TransHab discussed in Chapter 2 is not quite the same as the BEAM. While TransHab proved many of the aspects of inflatable module technology, there was still much work to be done. For example, although the NASA team had begun working on how to incorporate windows, putting holes in the module walls complicated the engineering of the skin so much that they left windows out of the proposed ISS module design entirely, although NASA filed a patent describing how windows could be fitted into a module. While details of how Bigelow has evolved the TransHab technology have not been published, one of the improvements was to include windows. To that end, Bigelow engineers conducted pressurization tests of different module designs, often deliberately pushing designs beyond their limits to establish safety margins. These tests often resulted in the module exploding. Fortunately, the team followed NASA's experience by conducting the pressurization tests under water, which muted the explosive force when the modules ripped apart. In one test, engineers decided to conduct a fill-to-fail test in open air and discovered why NASA hadn't when the test module exploded with such force that it shook the foundation of the building. In short, TransHab was a starting point for Bigelow's BEAMs, which is why it's worthwhile taking a look at the TransHab technology a little more closely.

If a TransHab had been deployed on the ISS, it would have provided a habitable volume (340 cubic meters) three times larger than a standard ISS (which has a total habitable volume of 388 cubic meters) module such as Destiny, which has a volume of 106 cubic meters. Not only would the ISS TransHab have provided facilities for sleeping, eating,

© Springer International Publishing Switzerland 2015
E. Seedhouse, *Bigelow Aerospace: Colonizing Space One Module at a Time*,
Springer Praxis Books, DOI 10.1007/978-3-319-05197-0_3

cooking, personal hygiene, exercise, entertainment, storage, *and* a radiation storm shelter, it would also have helped develop, test, and prove technologies necessary for interplanetary missions. We're not sure what the ISS crew would have done with all the extra space, but you can be sure they would have enjoyed living in it for a number of reasons. For one thing, a TransHab doesn't vibrate like the aluminum-shelled exterior of the rest of the ISS. The crew would also have appreciated TransHab's cavernous dimensions (7.0 meters long and 8.2 meters in diameter) and the open interior plan, which makes a confined volume psychologically beneficial to a crew in long-duration missions—an important factor considering the intense work and stressful environment astronauts face.

TRANSHAB NUTS AND BOLTS

TransHab (see sidebar and Figure 3.1) was a habitation module designed to support a six-member crew for long-duration stays in space. Originally designed to support a Mars mission, the concept was modified to support habitation needs on board the ISS. Functionally (see sidebars), it was designed to provide facilities for sleeping, eating, cooking, personal hygiene, exercise, entertainment, storage, and a radiation storm shelter.

TransHab Features

- An inflatable module comprising a structural core and an inflatable shell.
- In the launch configuration, the wall thickness of the inflatable shell would have been collapsed by vacuum and folded around the structural core.
- On deployment, the wall thickness of the inflatable shell would have been inflated around the structural core.
- Removable shelves were arranged interior to the structural core in the launch configuration.
- The core included at least one longeron[1] that, with the shelves, comprised the rigid, lightweight load-bearing structure of the module during launch.
- Removable shelves were detachable, which meant they could have been rearranged to provide a module interior suitable for human habitation and work.

---

[1]   A longeron is a thin strip of material to which the skin of an aircraft or spacecraft is fastened. The longerons run span-wise and attach between the ribs. The primary function is to transfer bending loads onto the ribs.

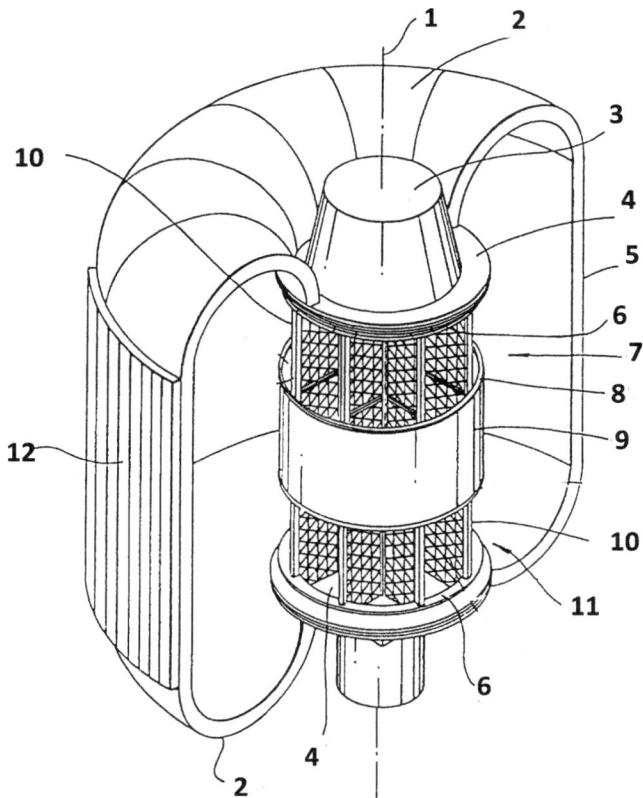

3.1 Partial cutaway isometric view of deployed module. Key: 1 = Longitudinal axis, 2 = Semitoroidal ends, 3 = Airlock, 4 = End plate, 5 = Circular cylinder, 6 = End ring, 7 = Structural core, 8 = Body ring, 9 = Water tank, 10 = Longeron, 11 = Shelf, 12 = Radiator. Inventors: Jasen Raboin, Gerard Valle, Gregg Edeen, Horacio De La Fuente, William Schneider, Gary Spexarth, Shalini Gupta Pandya, Christopher Johnson. Patent US6547189 B1. Courtesy: NASA/US

TransHab Advantages

- TransHab combined the advantages offered by and limited the drawbacks of pre-assembled modules.
- The habitat was lightweight, collapsible, and compact prior to and during launch.
- It could be enlarged to provide capacious volume for human habitation.
- Habitat minimized the number of parts to be assembled in orbit as well as the number of space walks to complete assembly.
- TransHab could be launched on several existing launch vehicles.
- Most components and system interfaces could be tested and calibrated prior to launch.
- Could easily be converted from launch configuration to deployed configuration.
- Utilized several key components for multiple functions, thereby providing mass and volume efficiency to the module.

TransHab Functions

- Individual private crew quarters for six astronauts
- A galley and wardroom
- Crew health care system
- Personal hygiene facilities
- Personal and general storage
- 300+ cubic meters of pressurized volume
- Internal pressure of 760 torrs (14.7 psia)
- Earth-viewing windows
- Secondary structure to support equipment and human systems
- Environmental control and life support
- Communications: human to human, human to machine/system
- Command and data handling
- Transition tunnel for crew and equipment into the ISS
- Lighting
- External structural interface to other pressurized modules
- External utility interfaces to the ISS
- Survive 10 years in space environment

TransHab at its most basic comprised a structural core and an inflatable shell attached to the structural core. In its launch configuration, the inflatable shell would have been collapsed by vacuum and folded around the structural core, and the module would have been loaded into the payload bay of a launch vehicle. Once in orbit, the module would have been deployed and the inflatable shell inflated. In its deployed configuration, the thickness of the inflatable shell automatically expands from its collapsed state to full thickness, and the inflatable shell is inflated around the structural core. Removable shelves are then arranged interior to the structural core in the launch configuration and can be rearranged to provide whatever interior the client requires. Before describing each level, it's useful to understand the function of each of the elements identified in the diagram. We'll begin with the structural core.

Structural core

TransHab's structural core (Figure 3.2) is cylindrical with a longitudinal axis and includes at least one longeron, one body ring, two endplates, and two end rings. The longerons extend parallel to the longitudinal axis of the cylindrical shape and are fixed to one of the

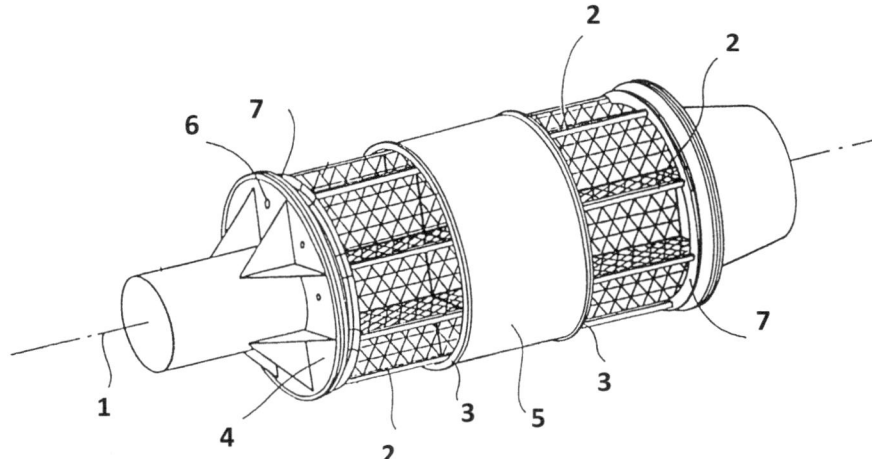

3.2 Partial cutaway view of structural core. Key: 1 = Longitudinal axis, 2 = Longeron, 3 = Body Ring, 4 = Endplate, 5 = Structural Core, 6 = Endplate pass-through hole, 7 = End Ring. Inventors: Jasen Raboin, Gerard Valle, Gregg Edeen, Horacio De La Fuente, William Schneider, Gary Spexarth, Shalini Gupta Pandya, Christopher Johnson. Patent US6547189 B1. Courtesy: NASA/US

circular end plates. Each endplate includes sealed pass-through holes that accommodate utility and umbilical conduits for plumbing, power, and data. The rectangular longerons, which extend along the periphery of the structural core's cylindrical shape, are fixed so the inner surface faces the longitudinal axis and the outer surface faces away from the longitudinal axis. The body rings you can see in the diagram are rectangular and are attached to each longeron, meaning the inner surface of each body ring abuts the outer surface of each longeron.

The structural core includes a cylindrical water tank extending between two body rings and around each longeron so the inner surface of the water tank abuts the outer surface of each longeron. The water tank is attached to two adjacent body rings and can be configured to include multiple independent water reservoirs. The reason for this arrangement is to provide radiation protection to the enclosed area—an advantage over ISS modules. Another feature of the structural core is removable shelves, each constructed in an isogrid pattern. In the launch configuration, the shelves would have been located within the cylindrical shape of module interior to the longerons. The shelving arrangement is very versatile, since each shelf-to-shelf attachment is able to mate with all other shelf-to-shelf attachments. Each shelf also includes plenty of regularly spaced attachment points, meaning the client can customize the habitat to whatever configuration is required.

A feature depicted in Figure 3.1 is the floor segment, constructed from a flexible material, such as corrugated graphite-epoxy sheet. In the launch configuration, the flexibility of the floor segment (in one direction but not the other) would have allowed each floor segment to be partially folded onto itself, allowing it to lie along the periphery of the cylindrical shape. Once deployed, the stiffness of the floor segments in the other direction would have allowed each floor segment to support weight and serve as flooring. The structural core also includes an airlock and several support system structures. Each of the two end rings is attached by means of welding or bolting at its end ring lower surface to the end of each longeron, meaning each longeron extends from the end ring lower surface of one of the two end rings to the end ring lower surface of the other end ring.

Inflatable shell

The shell (Figure 3.3) was a multilayer fabric construction that would have been folded around the structural core for compact packaging for launch and then inflated into the deployed configuration once on orbit. In the launch configuration, the inflatable shell would have been collapsed by vacuum and then deflated, collapsed, and folded around the structural core. Fully inflated, the shape of the inflatable shell would have resembled a circular cylinder with semi-toroidal ends. From inside to outside, the inflatable shell comprises an inner liner, multiple alternating layers of bladders and bleeder cloths, a structural restraint, a meteorite shield assembly, and an outer liner. You can think of the inner liner as TransHab's inside wall. The non-flammable inner liner would have protected the bladders from damage in the event something happened inside the module, but the liner also provided a means of attachment for the shape rings. As you can see in Figure 3.4, the shape ring has a circular toroidal shape, the outer diameter of which is sized slightly larger than

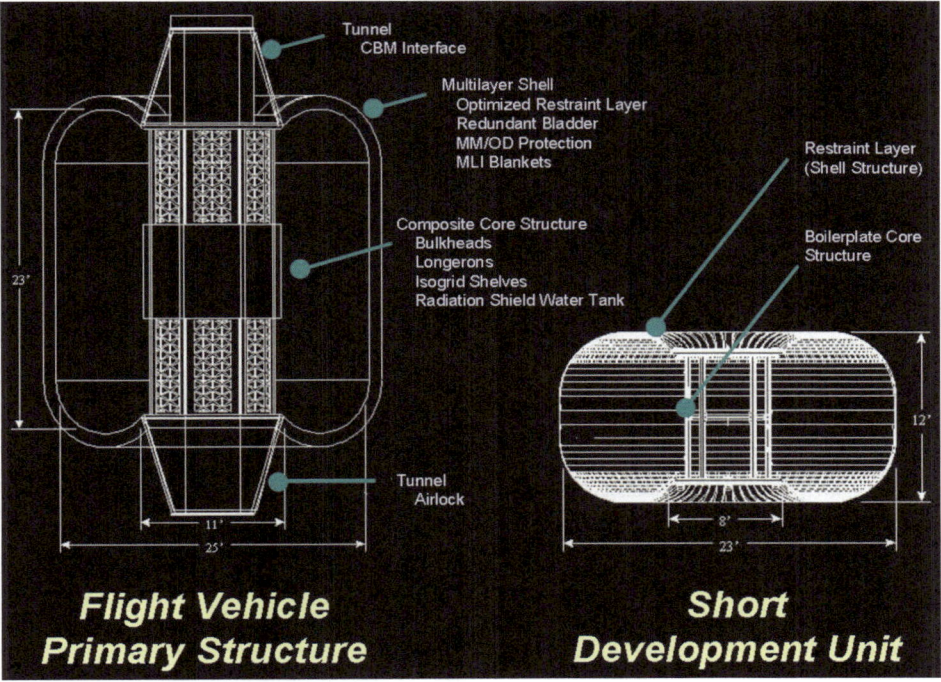

Tunnel
CBM Interface

Multilayer Shell
Optimized Restraint Layer
Redundant Bladder
MM/OD Protection
MLI Blankets

Restraint Layer
(Shell Structure)

Composite Core Structure
Bulkheads
Longerons
Isogrid Shelves
Radiation Shield Water Tank

Boilerplate Core
Structure

23'

Tunnel
Airlock

11'

25'

12'

8'

23'

**Flight Vehicle
Primary Structure**

**Short
Development Unit**

3.3   Flight vehicle structure showing tunnel, airlock, and multilayer shell. Courtesy: NASA

the cross-sectional diameter of the inner liner. This means that when the module was inflated, each shape ring would have abutted the inner liner and maintained a contact force radially outward on the layers of the shell.

Bladders

The arrangement of the bladders provided a redundant primary gas containment mechanism for the shell. The bladders were sealed together to form a thick single bladder just before the inner circumference of the semi-toroidal ends. Between each consecutive pair of bladders was a bleeder cloth—a lightweight, porous, felt-like material, the function of which was to prevent contact between bladders, thereby eliminating abrasion and providing a cavity between each bladder. This design also provided a cavity between adjoining bladders and allowed the pressure of each individual bladder to be monitored to aid in the identification and location of leaks. Another key feature of the bleeder cloths was their porosity, which enabled the equalization of pressure within each cavity. And, if need be, one or more bleeder cloth cavities could have been mechanically evacuated so that the lost gas from a bladder interior leak could have been recaptured and pumped back into the module. Bleeder cloth cavities could also be filled with liquid to provide added radiation protection (in the case of a Mars transit habitat, for example).

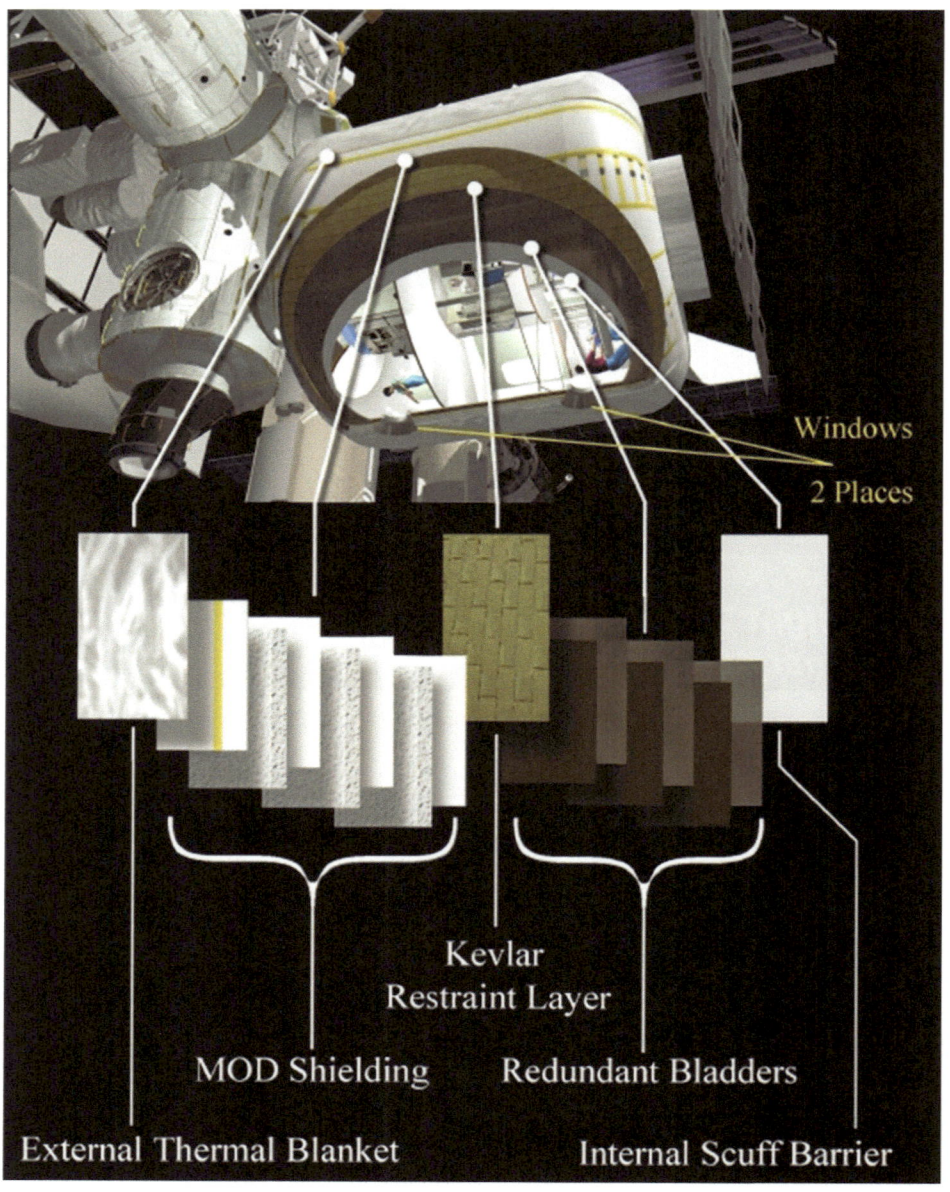

3.4 Exploded view of TransHab layers. Courtesy: NASA

Layers

The primary structure of TransHab was the structural restraint, constructed from Kevlar or Vectran. This layer served to separate the inner liner, bladders, and bleeder cloths from the shield assembly and outer liner of the inflatable shell. But perhaps the toughest layer was the shield assembly, comprising several bumper layers, spacing layers, and adhesive. Each of these layers-within-a-layer had a job to do to protect astronauts from meteorite and/or orbital impact. For example, the bumper layers would have fragmented and vaporized incoming particles and, because of the extreme heat energy generated in the resulting particle vapor cloud, each bumper layer was coated with an ablative energy-absorbing adhesive, which also acted as an ablator for added shielding. The bumper layers were separated by a spacing layer composed of open cell foam, and each spacing layer included gaps located to create hinge lines on the shell to enable folding of the shell for launch. Since each spacing layer comprised open cell foam, the shell could easily be collapsed by vacuum evacuation for launch packaging. Then, when exposed to the vacuum of space, the spacing layer would have returned to its original thickness.

TransHab Layer by Layer

TransHab comprised several layers (Figure 3.4), including an inner liner, a bladder, a structural restraint layer, a micrometeorite/orbital debris (MMOD) protection system, a thermal control layer, and an atomic oxygen protective layer. The inner liner served to protect the bladder from internal hazards by providing a durable flame- and puncture-resistant barrier. The reason for the inner layer having to provide such robust protection for the bladder was because the bladder was the primary gas barrier layer that maintained the air volume: if this failed, then an explosive decompression event would have put an end to the mission. Because the bladder was such an important layer, the engineers thought long and hard about what the best design should be. It was a tough task because the bladder had to be flexible and durable during manufacturing, assembly, and folding, and deployable under extreme temperatures. The easiest design was a single-layer bladder because a single layer would have been easier to design and manufacture but it wouldn't have had the benefit of redundancy. But a double/multiple-layer bladder design was complicated because it required ports between the bladder layers to allow venting on ascent and monitoring of pressure while on orbit.

The outermost layer, which comprised a lightweight multilayer insulating blanket, also provided an impermeable membrane to enable the vacuum compaction of the spacing layers prior to packaging. Attached to the exterior was a radiator, designed to be folded along with the shell. The shell was attached to the structural core at each end ring and the layers ended in a deadman. This deadman, whose function is akin to an anchor attached to the end ring by way of a ring-shaped retainer, was securely attached to the primary ring component by dead-man retainer bolts equally spaced about the circumference of the deadman retainer. Adding to the structural integrity of the shell were the attachment rings, each connected to the structural restraint at the inner circumference of each semi-toroidal end by folding and stitching the structural restraint around the attachment ring. Each attachment ring included an outer surface, an inner surface, a top surface, and a bottom surface. In addition to the attachment rings, TransHab featured end rings, attached to each seal ring by means similar to the bolting used to secure the deadman retainer. The seal rings, meanwhile, helped secure retain the inner liner and, while we're on the subject of seals, it's time to discuss the inclusion of windows, which presented engineers with all sorts of headaches.

Windows

Simply cutting a hole in the layers of the module wouldn't do because this would have dramatically reduced the tensional loading on the layers and also reduced the circumferential integrity of the structure. The engineers also had to contend with the problem of distributing tensional loads equally, which was no easy task when dealing with a combination of flexible and rigid materials. Then there were the problems associated with bending moments on the frame, non-axial stresses on the connecting mechanisms, and making sure no part of the structure was under-loaded or over-loaded as a result of the inclusion of two or three windows. All in all, the window problem required some ingenious out-of-the-box thinking, and that's exactly what the NASA engineers did. By implementing strategically positioned connecting mechanisms, adjusting tensions on the straps, and using a system of rollers to ensure loads were distributed equally, the window problem was solved, as described here.

Integrating a rigid structure in the flexible wall of an inflatable module was a major challenge to TransHab engineers because these "structural pass throughs", or openings, had to be formed through the bladder, the inner and outer protective layers, *and* the flexible restraint layer. To prevent the escape of gas from the module through such an opening, there had to be some way to attach the bladder to the edges of the window frame. This meant the longitudinal and circumferential straps (Figure 3.5) adjacent to a window had to be terminated to form the opening in which the window frame would be mounted. To do this without compromising the structural integrity of the module wasn't easy, since it meant there had to be some way to attach the looped end portions of the straps to portions of the window frame while maintaining tensional load. There was also the issue of integrating the rigid and flexible components such as the bladder, restraint layer, and protective backing materials as the module was unfolded and expanded during its inflation and deployment. This was a concern during and after deployment because differential stresses and different reaction to stresses by rigid and flexible components could have resulted in excessive strain on the flexible components. In short, tensional forces exerted on flexible straps could have resulted in stretching of the straps, whereas the same forces exerted on a rigid panel would have produced little deformation. So, the engineers had to figure out how to reduce differential stresses on the flexible straps as well as making sure the forces were evenly distributed.

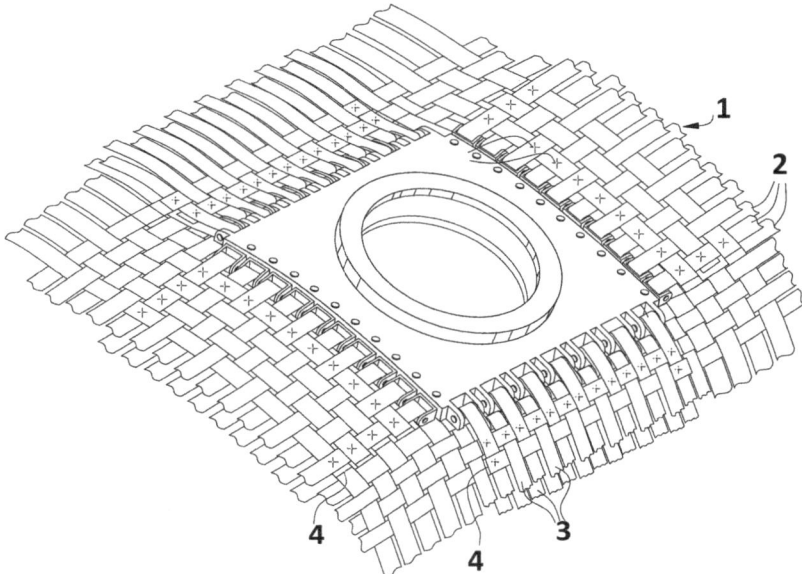

3.5 Perspective of a rigid panel as viewed from the outside of a module showing circumferential and longitudinal straps of the restraint layer. Key: 1 = Restraint layer, 2 = Reinforcing straps, 3 = Load-bearing straps, 4 = Supplementary straps. Inventors: Christopher Johnson, Ross Patterson, Gary Spexarth. Patent 7509774 B1. Courtesy: NASA/US

You can get an idea of how the engineers resolved the problem in Figure 3.5. The two key features here are the flexible, load-bearing straps and the window. The window had several connecting points to which load-bearing straps could be attached, which meant tensional loads on the straps could be divided between the straps in a contiguous alignment. One option was to have a curved window with a curvature corresponding to the flexible wall whereas another option was to use a flat window. As you can see in the inset to Figure 3.6, the module included several window frame assemblies mounted in the module flexible outer shell. The window frames were mounted in and supported by portions of the restraint layer and connected at the edges to the straps. The circumferential and longitudinal straps intersecting the frame terminated adjacent to the frame to form the rectangular opening in which the frame is supported, and the end portions of the intersecting straps were folded back to form loops in the end portions of the straps at their juncture with the frame. Why the seemingly haphazard location of windows? Well, the reason the window frames are not aligned with one another along either the longitudinal or circumferential axes is because this avoided making multiple, successive breaks in the straps, thereby preserving structural integrity. Another feature key to reducing structural loads was the use of a window frame curved along its length—a choice that reduced differential and torsional loads on the window frame.

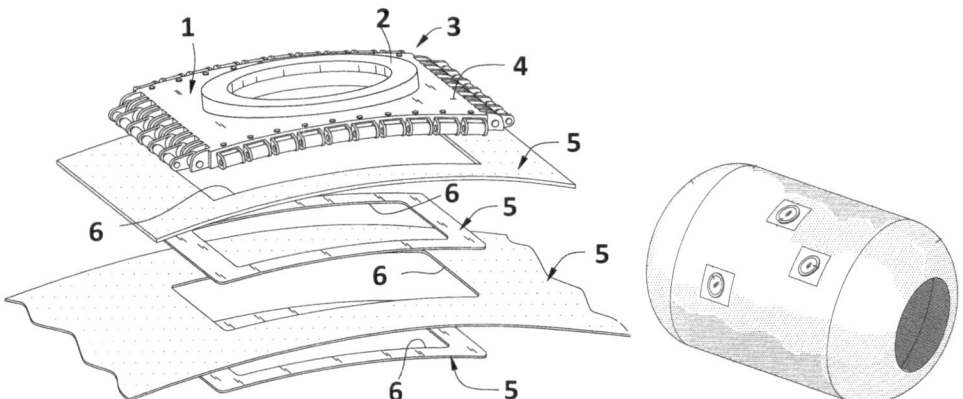

3.6 Exploded isometric representation of panel structure, showing components extending along-side the inner side of the panel structure. Key: 1 = Window frame, 2 = Window seat, 3 = Frame assembly, 4 = Outer surface window frame, 5 = Window frame layers, 6 = Rectangular openings of window frame layers. Inventors: Christopher Johnson, Ross Patterson, Gary Spexarth. Patent 7509774 B1. Courtesy: NASA/US

Connecting mechanisms

Looking closer at the connecting mechanisms (Figure 3.5), you can see outer and inner connecting mechanisms mounted on the frame end portion. Each outer connecting mechanism included an outer roller structure and each inner connecting mechanism included an inner roller structure, each adapted to receive the looped end portion of one of the circumferential straps. The integrated assembly of the roller structures was supported by lugs that projected outward from the frame end portion and were able to rotate about the axis. The staggered array of connecting mechanisms looks complicated, but the way they were arranged meant they were well suited for receiving and distributing the circumferential loads. And, on the subject of circumferential loads, it is necessary to mention the straps, because this feature provided important structural benefits: the circumferential straps extended perpendicularly outwardly from the adjacent frame end portions to which they were connected, and the longitudinal straps extended perpendicularly outwardly from the respective frame side portions to which they were respectively connected, and perpendicularly to the circumferential straps. In contrast, the circumferential straps were in a side-by-side orientation, and the longitudinal straps were mutually spaced—a feature that also provided important structural benefits. How? We'll get to that shortly, but first a description of the window frame.

Getting into the guts of the design in Figure 3.7, we can see (for clarity, sections of straps are omitted in Figure A) the circumferential straps were connected to the end portions of the window frame by outer connecting mechanisms and inner connecting

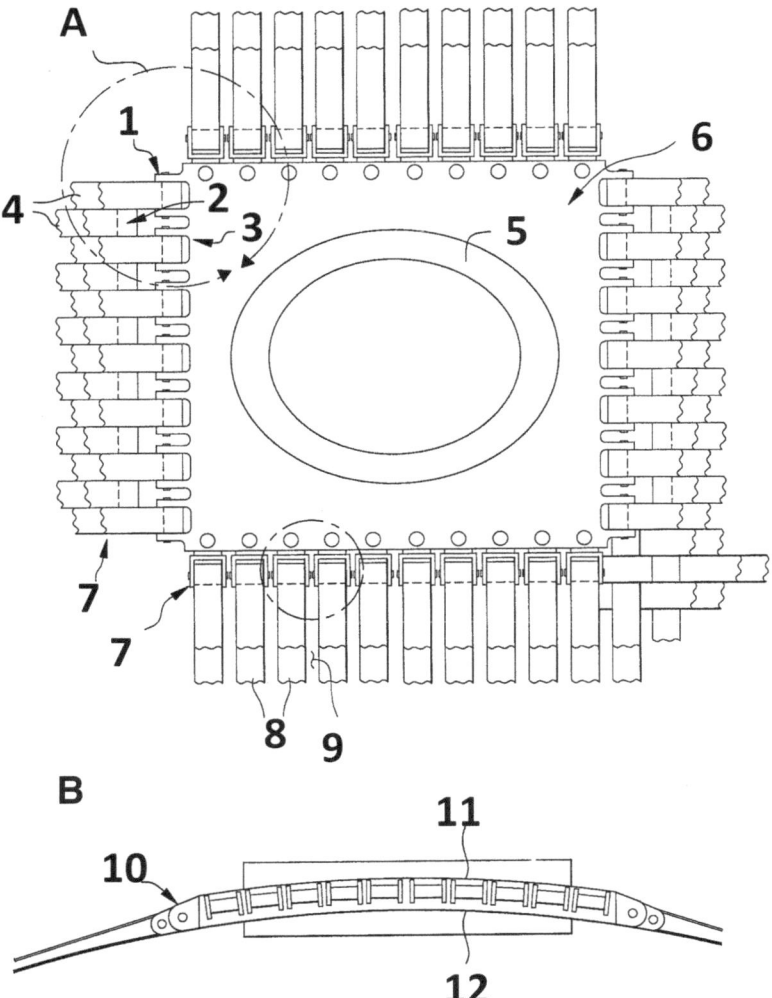

3.7 **(A)** Plan view of panel structure viewed from outside. **(B)** Side view of panel and circumferentially extending straps of structure shown in (A). Key: 1 = Inner connecting mechanisms, 2 = Outer connecting mechanisms, 3 = Integrated assembly, 4 = Load-bearing straps, 5 = Window seat, 6 = Window frame, 7 = Connecting mechanisms, 8 = Longitudinal straps, 9 = Gap, 10 = First end portion, 11 = Outer surface, 12 = Inner surface. Inventors: Christopher Johnson, Ross Patterson, Gary Spexarth. Patent 7509774 B1. Courtesy: NASA/US

mechanisms mounted on frame end portions, forming staggered arrays of connecting mechanisms extending along the lengths of end portions. The longitudinal straps were connected to the side portions of the window frame by the connecting mechanisms mounted on the window frame side portions and these straps were supported by inner connecting mechanisms parallel and laterally spaced from one another (indicated by the number 9). You will also notice loops are formed in the ends of the circumferential and longitudinal straps where they join the frame so the straps could be connected to the frame and also extend along the circumferential path.

The descriptions of the engineering challenges so far have been from the outside, so let's take a look at the window frame as if we are inside the module. As you can see in Figure 3.7, the arrangement of connecting mechanisms and fastenings is similar to the arrangement for the exterior, but there are differences such as the bladder and bladder connection arrangement. The bladder was positioned inwardly of the restraint layer webbing formed by straps and was connected to the inner surface of the frame end portion but, to protect the bladder from possible protrusion into gaps formed between the connecting mechanisms, a layer of stiff felt material was fastened over the bladder. The felt layer together with the arrangement for connecting the straps, the connecting mechanisms, and protective felt buffer layer combined to provide a smooth transition of the restraint layer and bladder into the frame.

The net effect of all this engineering was a safe means for integrating rigid structures into the flexible wall of an inflatable structure and ensuring a relatively low-stress interface between the rigid and non-rigid components. The design also prevented damage to the flexible layers in the shell, and accommodated the differential stresses and differing reaction to stresses by rigid and flexible components during and after deployment, as the shell expanded.

Launch restraint

With all these problems solved, the next challenge was to ensure the module could be launched in its launch restraint, the element that maintained the inflatable shell in its folded arrangement around the structural core. The launch restraint had to be designed to be easily releasable so an astronaut could detach the restraint once the module was ready to be deployed. One way to do this was to devise a zip cord mechanism released by a simple pulling motion, which, once activated, unwound from around the inflatable shell.

The carrier was a lightweight mechanism that would have been used to transport the structural core and inflatable shell in the launch configuration. In addition to providing transportation, the carrier would have isolated the core from the large bending loads generated during the launch. In preparation for launch, the inflatable shell, including the spacing layers of the shield assembly, would have been shrunk by vacuum, and the shell deflated, collapsed, and folded around the structural core and fairing, with the launch restraint maintaining the shell in its folded arrangement around the core. Once assembled in its launch configuration, the module would have been positioned within the launch vehicle's payload bay and the vehicle would have lifted off to Earth orbit. During the ascent to orbit, the launch loads on the module would have been absorbed by the configuration of the longerons, body rings, endplates, end rings, shelves, and airlock. To prevent buckling during the ascent, a body ring would have been attached at each longeron buckling mode node location, thereby imparting sufficient lateral stiffness to the longerons.

Deployment

Once stabilized in orbit, the module would have been ready to be deployed. First, the launch restraint would have been unfastened from around the inflatable shell, after which the shell would have expanded into its deployed shape, thanks partially due to the vacuum of space as well as the general structural shape provided by the spacing layer of the inflatable shell. With the restraint released, the open cell foam would have returned naturally to its original thickness, which would have helped expansion. To complete the inflation, the shell would have been inflated by pumping a gas into the interior thanks to an inflation system attached to one of the endplates. Once the atmospheric pressure was sufficient for human habitation, the end rings between the structural core and the shell would have been sealed, thereby preventing gas from leaking. Once that was done, the shape rings would have been inflated and placed on their hook and pile systems on the inner liner. The shape rings would have helped maintain the shape of the shell and provide mounting locations for the shelves.

Floors and floor segments

With the module inflated, it would have been time to configure the habitat for operations, which would have meant installing the floors. The floor segments were attached to struts immediately adjacent to each other on each body ring and, since each floor segment was partially folded onto itself between its corresponding struts, it would have been necessary to reconfigure the struts to lock the floors into place. Once the floors were in place, our astronaut construction workers could have gone to work configuring each level into quadrants or sections, depending on what TransHab was being used for. Part of the finishing touches would have included installing the shelves, which would have been detached from their stowage position. Once that was done, the astronauts could have taken a tour of their new inflatable home.

TRANSHAB LEVEL BY LEVEL

The ISS TransHab was divided into four functional levels: Levels 1, 2, and 3 for living space and Level 4 the connecting tunnel. Because TransHab was a prefabricated, packaged, and deployed habitat, the crew would have been required to perform set-up and outfitting activities to make it operational.

Levels 1 and 3 were 2.4 meters tall at the central core and Level 2 was two meters tall at the core. From inside bulkhead to inside bulkhead, TransHab was seven meters long (this length doesn't include the Level 4 pressurized tunnel). After docking with the station, TransHab would have been berthed, deployed, and inflated to its internal operating pressure of 14.7 psia. Following inflation, systems would have been activated to condition the environment for crew entry and outfitting.

Level 1

This was the galley/wardroom and soft stowage area (Figures 3.8 and 3.9). One of the unique features of this level was the two-storey clerestory located above the wardroom area. The clerestory, or "great room", was included in response to the psychological need for an open space for the crew. When you're cooped up in a can—or a series of cans in the case of the ISS—for months at a time, it's easy for crew morale and productivity to suffer, even if you happen to be a highly trained astronaut. So, to alleviate this problem, designers, together with astronaut input, decided it would be a good idea to include the clerestory, sporting a galley rack, refrigerator/freezer racks, and a large wardroom table, designed to gather up to 12 crewmembers. The soft stowage area consisted of the stowage array system and a hand wash.

Level 2

This was the upper level of the Level 1 clerestory, the mechanical room, and the crew quarters (CQ). The CQ area (Figures 3.10 and 3.11) had six crew quarters and a central passageway located within the second-level central core structure and radiation shield water tanks. Thanks to a mezzanine level, the mechanical room was located outside the

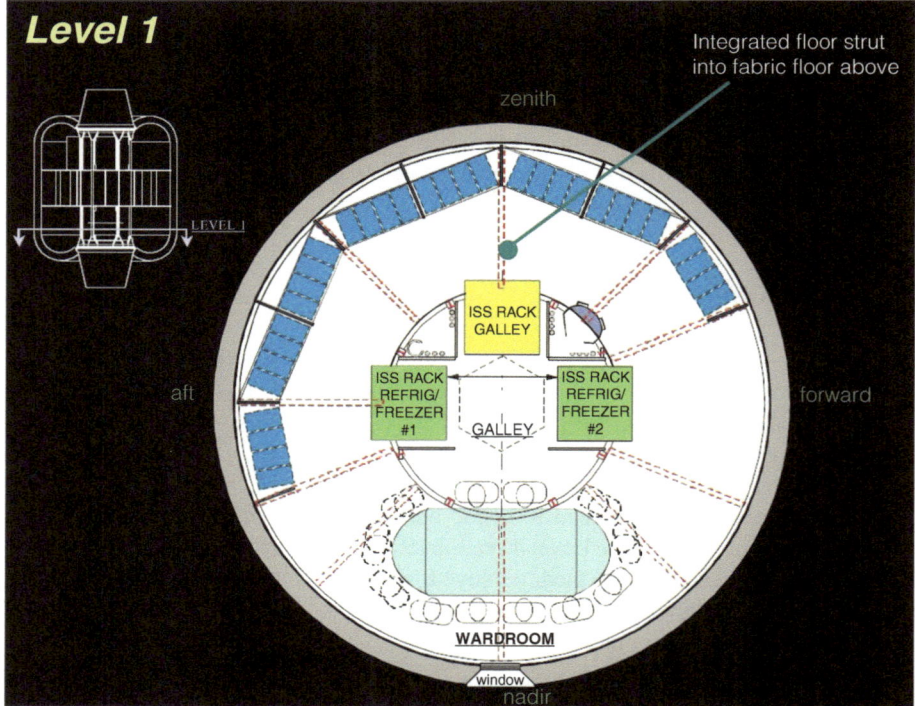

3.8 TransHab cross-section of Level 1. Courtesy: NASA

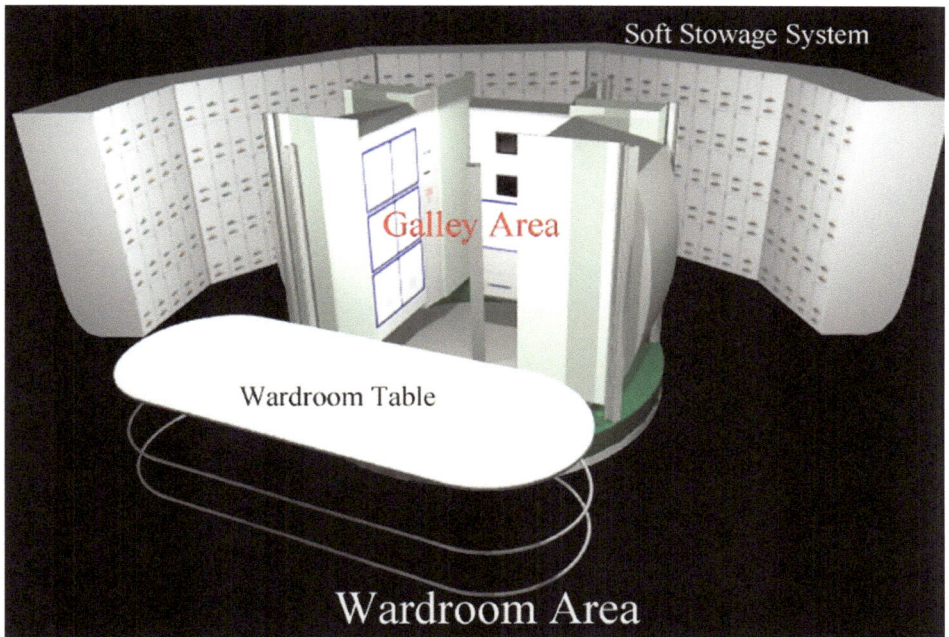

3.9 TransHab interior of Level 1. Courtesy: NASA

core structure and used only half the floor space. Acoustically and visually isolated from the rest of TransHab, the mechanical room was designed to house the Environmental Control and Life-Support System (ECLSS), and the power and avionics equipment. Openings in the mechanical room floor and ceiling along the shell wall provided return airflow from Levels 1 and 3. The open and flexible design of this level meant equipment was easily accessible; equipment was integrated onto the shelves that would have been placed into the core for launch, and the equipment shelves would have been moved to their final location once the habitat was inflated.

One feature that would have particularly appealed to crews was the 6.35-centimeter-thick radiation shield water jacket that surrounded the CQ. Access to this area was from Level 1 (below) or Level 3 (above), via the central passageway. Each of the CQ was a spacious 2.27 cubic meters (213.4 centimeters high), which is larger than the ISS rack-based CQ. Each CQ would have had personal stowage, a personal workstation, sleep restraint, and integrated air, light, data, and power. A nice touch was the acoustic wall panels designed for change-out—a capability that would have allowed new crewmembers to bring personalized panels to decorate their crew quarter according to personal taste. As with so many design aspects of TransHab, this concept was based on the outcome of studies that investigated factors impacting crew morale and productivity.

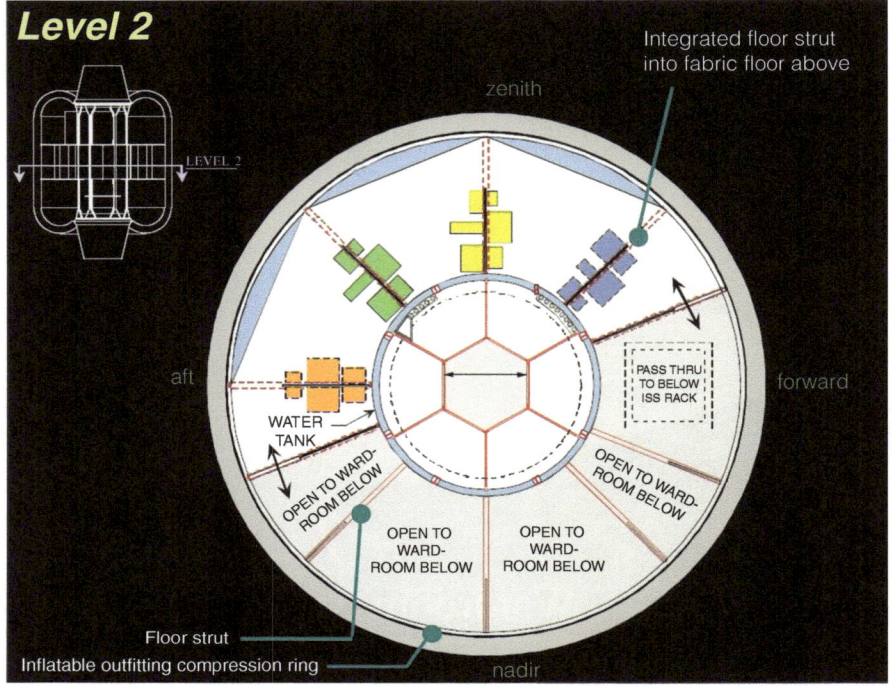

3.10   TransHab cross-section of Level 2. Courtesy: NASA

Level 3

Level 3 (Figures 3.12 and 3.13) was the crew health care and soft stowage area, housing the Crew Health Care System (CHeCS) racks, a Full Body Cleansing Compartment, changing area, exercise equipment (treadmill and ergometer), a screened area for medical exams and conferences, and an Earth-viewing window. Also found here was a soft stowage area identical to Level 1. To save the crew valuable time, the exercise equipment would have been in a permanently deployed position and placed near the window to allow the crew to view Earth during their two-to-three-hour exercise training sessions. Two equipment shelves would have been placed on the floor struts as exercise equipment mounting platforms and structural integration. Four movable partitions would have provided visual screening of crewmembers for pre- and post-full-body cleansing activities and private medical exams at the CHeCS rack.

Level 4

Level 4 was the pressurized tunnel that featured two ISS common hatches, and avionics and power equipment. Functionally, it was designed to provide a transition between Node

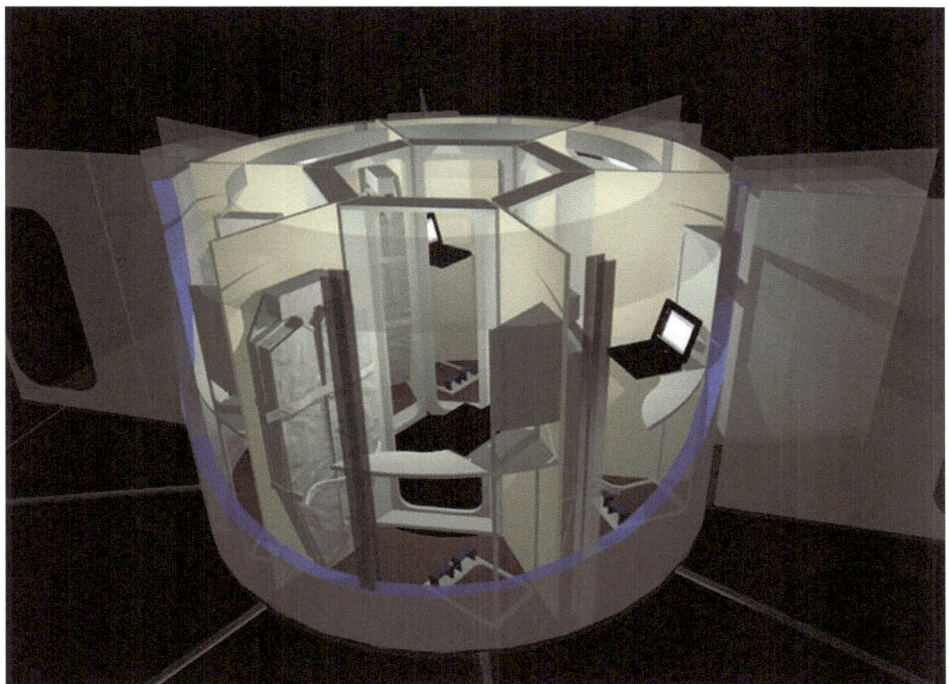

3.11 TransHab interior of Level 2. Courtesy: NASA

3 and TransHab, house-critical equipment required during inflation, and provide a structural connection to the ISS. During launch, it would have been the only pressurized volume in TransHab until inflation.

TESTING

So just how strong and safe is an inflatable space structure? Very strong, as it turns out. TransHab was subjected to rigorous testing and passed with flying colors, as described here. We'll begin with the restraint layer, which was TransHab's structural load-bearing layer designed to support the bladder and carry the load induced by the internal pressure. The challenge when designing the restraint layer was coming up with a design that could carry a high load—a function of the large volume and high pressure inside TransHab. That pressure was internal pressure that imparts stress as hoop stress and longitudinal stress. Because TransHab geometry comprised a cylindrical central section and half-toroid-shaped end-caps, engineers focused on the cylinder and toroid vessel shapes when trying to determine the restraint layer's strength requirement. Once they had devised what they reckoned was a strong structure, it was time to test. The metallic pressure vessels used on the ISS are designed to a factor of safety (FOS) of 2.0 but, due to the uncertainties when

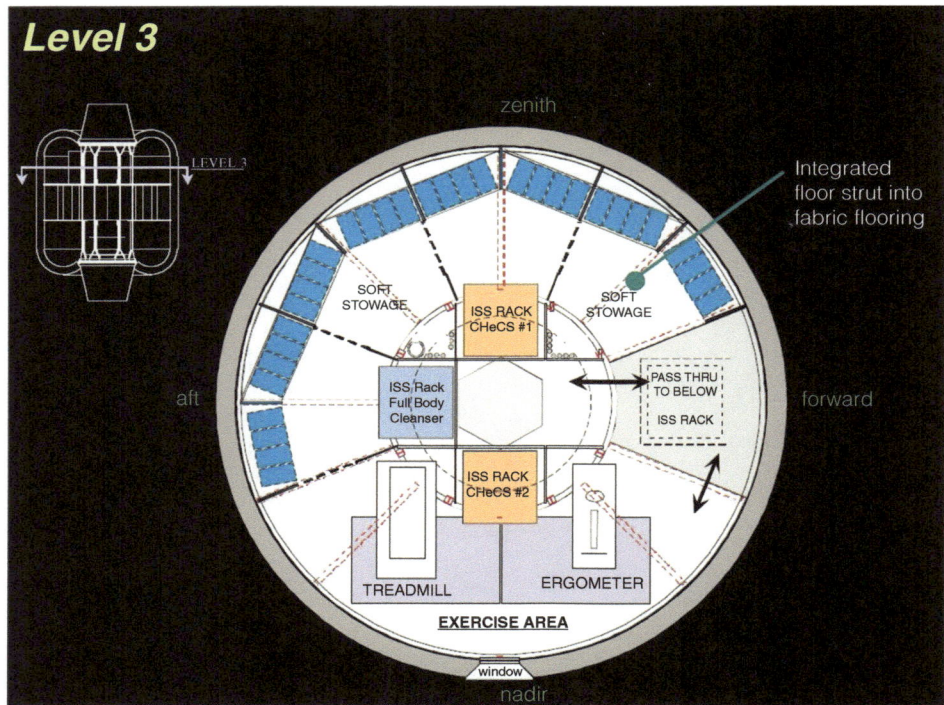

3.12   TransHab cross-section of Level 3. Courtesy: NASA

using fabrics, the Federal Aviation Administration (FAA) requires all inflatable airships be designed to FOS of 4.0, which is why the TransHab team decided to impose the same requirement on the restraint layer. One problem they encountered in meeting the strength challenge was the seams. Structural seams in high-strength webbings have seam efficiencies between 80% and 90%, which means the seams are 10% to 20% weaker than the general fabric strength. The solution to make up this loss of strength is to add 10% to 20% to the strength requirement of the fabric but, for the TransHab engineers, it wasn't that simple because there can be significant variance depending on material strength, width, weave style, and seam type. The engineers also had to find a strong—and affordable—fabric that could meet the FOS requirement. A common high-strength fabric that nearly made the grade was PBO (phenylene benzobisoxazole), a material with almost twice the tensile strength of Kevlar or Vectran. The problem was PBO was expensive and difficult to prepare, so the engineers turned to Spectra, a fiber spun from a solution of Ultra High Molecular Weight Polyethylene (UHMWPE). Spectra had adequate tensile strength, but its brittleness at low temperature eliminated it as a viable material. Finally, the TransHab program selected Kevlar thanks to the material's significant flight history, well-known properties, and low cost.

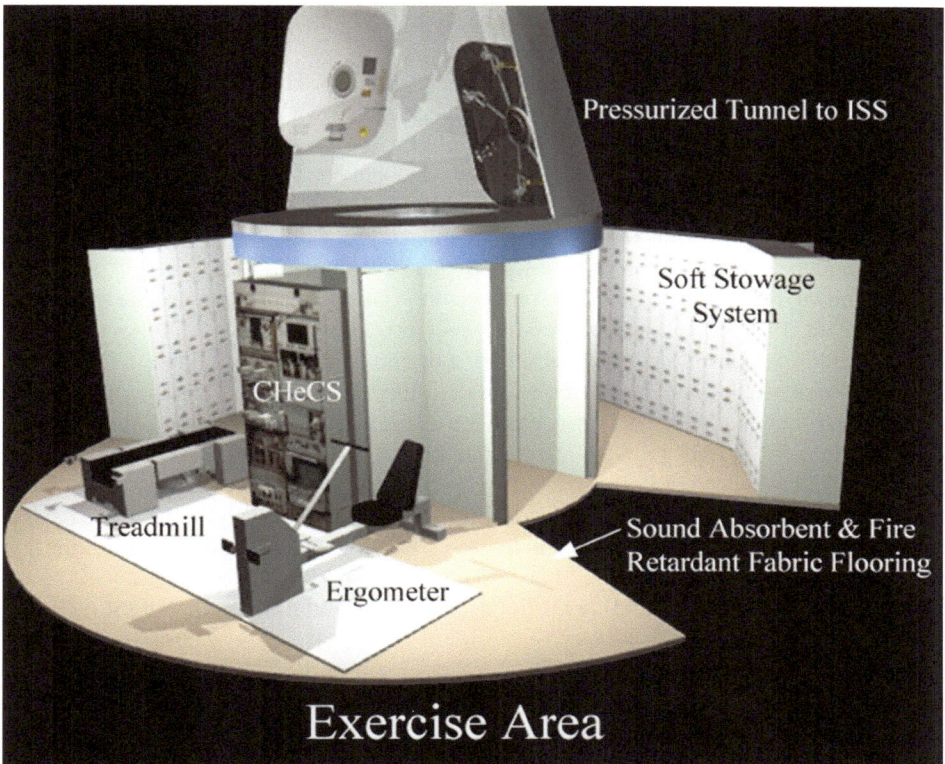

3.13 TransHab interior of Level 3. Courtesy: NASA

Once the TransHab team had selected the material, they had figure out a way to build the restraint layer to test the various materials and structural loads. To do this, they built full-scale Shell Development Units (SDUs) and used them as test articles to assess different fabric-style options such as multiple layers, multilayer fabric, wide or narrow webbing, or multiple fabric layers. Constructing the multiple-layer option required sewing several layers of Kevlar together to achieve the strength necessary to withstand the pressure loads. One problem with this approach was that it was very difficult to achieve proper load-sharing throughout the layers. Another problem was the reduction in strength when cutting gore[1] sections of the material, which was necessary when designing curved geometrical shapes. The multi-fabric option also needed to be cut into gore patterns to achieve the

---

[1]   A gore is a sector of a curved surface that lies between two close lines of longitude on a globe and may be flattened to a plane surface with little distortion. The term has been extended to include similarly shaped pieces such as the panels of a hot-air balloon or parachute, or the triangular insert that allows extra movement in a garment.

toroidal end-caps, which meant it shared the same load-sharing and strength concerns for the multiple-layer option.

Because of the problems with the fabric options, the TransHab team decided to construct SDU-1 out of wide webbing, but this approach had its own drawbacks. First, there was uneven loading near gaps in the restraint layer created as a result of the geometry. The second drawback was that it wasn't possible to get adequate load-sharing across the webbing because, to form the inflated shape, the flat webbing had to assume a curved shape across the strap width. The wide-webbing design also suffered from the fact that it was difficult to manufacture because it proved very difficult to maneuver the heavy, dead weight of the webbing around to the sewing machines to produce the seams. Nevertheless, despite these snags, on May 28th, 1998, SDU-1 was successfully pressurized to more than 2.5 times operational pressure, thus demonstrating the capability of a design using wide webbing.

Because of all the problems encountered in testing the wide-webbing approach, SDU-2 was constructed using narrow (2.4-centimeter-wide) webbing. Due to the large number of narrow straps required, the SDU-2 restraint layer was woven by hand. Unfortunately, when the SDU-2 was first inflated, it was soon apparent that the weave was loose and not conforming to the intended geometry, resulting in the SDU-2 being disassembled and re-woven. The re-woven SDU-2 was inflated a second time and this time the restraint layer inflated to the correct geometrical shape. On September 12th, 1998, SDU-2 was successfully proof-tested to a FOS 4.0 at the Johnson Space Center (JSC), making it one of the highest loaded inflatables in history.

Shortly after the SDU-2 test, JSC began fabrication of the SDU-3 restraint layer. This layer was also woven by hand and contained over 16 kilometers of Kevlar webbing that was assembled using more than 26,000 hand stitches. The SDU-3 was a full-scale unit that sported a debris shield installed over the restraint layer and was used to demonstrate folding and deployment in a vacuum. In designing and building three SDUs, JSC proved that highly loaded, large-diameter inflatable structures could be manufactured to meet demanding design requirements, as well as be manufactured in much less time than conventional aluminum structures. The next challenge was to provide protection for TransHab.

Micrometeoroid/Orbital Debris (MMOD) protection system

Regardless of whether inflatable modules were used in low Earth orbit (LEO) or en route to Mars, lightweight, fully flexible shields were required to protect TransHab from meteorite and debris impact. One option was to add shielding after inflation, but this would be expensive and labor-intensive. A more desirable—and elegant—solution was to use a pre-integrated, deployable, fabric shield, ideally one that exceeded the performance of conventional shields. Although there were a number of types of MMOD shields available at the time, JSC was most interested in a pre-integrated compressible and deployable, multilayer, fabric shield—one that provided state-of-the-art hypervelocity impact protection. As described earlier, the TransHab shield consisted of several layers of Nextel ceramic fabric layers separated by open cell polyurethane foam, which provided a standoff distance between the Nextel layers. Prior to launch, the foam would have been vacuum compressed, to minimize volume and allow the shell to be folded. Once on orbit, the foam would have regained its original thickness due to the lack of differential pressure. Behind the alternating layers of Nextel and foam was a layer of Kevlar. As particles impacted the Nextel layers, they would be progressively shattered into smaller, slower particles over a larger area. With a suitably sized shield, by the time the particles reached the Kevlar rear wall, they would be small and slow enough to be stopped. To test this theory, the JSC team built a shell mock-up and performed hypervelocity impact testing (see sidebar) at JSC and the White Sands Test Facility. It was a very important series of tests because, if the debris shield couldn't stop the particles, then TransHab had no hope of surviving in LEO. The test engineers fired shot after shot after shot, but the shield held. Initially, the engineers, who loved blowing stuff up, were a little disappointed, until they realized what a breakthrough the shield's ruggedness represented.

Blowing Stuff Up

To test TransHab's ability to resist impacts by micrometeorites, engineers conducted hypervelocity impact tests, using differently sized particles at speeds ranging from 2.5 to 11 kilometers per second. By comparison, the speed of a bullet is around 1.2 kilometers per second. The first goal was achieved by building a typical shell lay-up and performing hypervelocity impact testing at JSC and the White Sands Test Facility. The 30.5-centimeter-thick orbital debris shield took shot after shot and refused to fail, exceeding all expectations. During follow-up tests, TransHab's shell survived impacts of a 1.7-centimeter aluminum sphere fired (at 0° and 45° impact angles) at a velocity of seven kilometers per second.

Thermal control layer

In addition to the MMOD shield, TransHab needed a thermal control system to maintain a habitable internal environment. One layer that came under particular scrutiny was the atomic oxygen protective layer, which was needed due to combat the corrosive effects of atomic oxygen (AO). Since all spacecraft in LEO are exposed to AO, Betaglass fabric is usually used to protect against AO damage. In the upper layers of the atmosphere (90–800 kilometers), atmospheric atoms, ions, and free radicals, most notably AO, play a major role in the corrosion of spacecraft. The concentration of AO depends on altitude (and solar activity) and, between 160 and 560 kilometers, the atmosphere consists of about 90% AO. Different materials resist corrosion in space differently. For example, aluminum is slowly eroded by AO, while gold and platinum are highly corrosion-resistant. To assess the corrosive effect of AO on more than 400 candidate spacecraft materials (including some used for TransHab), NASA devised the Materials on International Space Station Experiment (MISSE), which exposed a variety of materials to the LEO environment for up to four years[2] between August 2001 and August 2005. Following return to Earth, these materials were analyzed to determine which materials could withstand the space environment and which could be used in the design of future spacecraft. For the most part, the TransHab materials survived their exposure very well. Except for the foam and bladder materials, which showed evidence of heavy erosion, no AO reactivity was noted and the Kevlar and Nextel materials were intact.

Deployment system

Having tested the materials, the next challenge on the agenda was testing a system to restrain the folded shell layers prior to deployment. This system had to withstand the pressure build-up due to venting of the shell layers during ascent and it also had to fit into the rocket fairing as snugly as possible. One option was to use a deployment system external to the shell layers, which would have had the advantage of eliminating the problem of contamination (outgassing) as long as the external deployment system was removed and properly disposed. The alternative was to use a system integral to the shell layers, which was the approach used during the vacuum deployment test. JSC settled on an integral system as described earlier, comprising a series of deployment straps spanning every third deployment gore. When the test habitat was folded, every third gore would be pushed in towards the central core. Adjacent gores were folded over so that the ends of the deployment straps on one deployment gore lined up with the ends of the deployment straps on the next deployment gore. Deployment cords were tied to each end of the deployment straps and strung together like a daisy chain. The deployment straps formed multiple segmented rings that contained the folded assembly. Each set of daisy chains could be released from a single "cut" location using pyrotechnic guillotine cutters. The deployment system was successfully tested in a vacuum using the TransHab SDU-3 full-scale deployment test at vacuum.

Shell-to-core interface

After proving the deployment system, engineers considered the issue of the shell-to-core interface. This interface between the inflatable shell and the central core was the most critical area of the module because this was where the bladder had to maintain a leak-tight seal. As with the deployment system, engineers had design options, one of which was to design a conical compression ring that the restraint layer could wrap around. In this design, as the load in the restraint increased, the pressure load on the ring and attached interface also increased. One advantage of this type of attachment was that, while the load increased, the conical ring would react to the load in hoop tension. The downside to this design was the chance that there might not be good load-sharing across and around

---

[2]   MISSE-1 and -2 were deployed in August 2001 on Expedition 3 and were planned for a one-year exposure but, due to the delays following the *Columbia* accident, they were not retrieved until four years later during ISS Expedition 11 in August 2005.

the restraint layer because it could be compressed against the interface. This could damage the restraint layer, although no such issues were reported in the successful test of the SDU-1 that used this design.

Folding and vacuum deployment

Next, the engineers went to work on the folding and vacuum deployment test. This test would demonstrate the ability to assemble, package, and deploy the multilayer shell, a full-scale SDU-3. In December 1998, SDU-3 was folded in a vertical configuration using a series of four cables attached to each of the 21 gore interfaces. The 84 cables were attached to an overhead support fixture that supported the 4,500-kilogram shell weight and allowed each gore to be folded. To fold the 21 gores, every third gore was pushed in towards the central core, and each of the seven adjacent gores was folded over, after which the test article was deflated, transferring the shell weight to the overhead support structure. Kevlar webbing was then used to draw in the 14 gore-to-gore seam interfaces and SDU-3's deployment system was laced up daisy-chain fashion and tied off with pyrotechnic test cords. The test article was successfully folded with minimal ground support equipment and the final packaged diameter was small enough to fit in the Shuttle's cargo bay. Once the test article was folded and all pyrotechnic cutters were armed, JSC's cavernous thermal vacuum Chamber A was pumped down to approximately 27 torr. Then, the pyrotechnic cutters were fired, releasing the packaged shell, after which SDU-3 was re-inflated to ambient differential pressure.

TransHab's demise

By the end of 1998, the JSC team had proven the technologies required to design, fabricate, and utilize an inflatable module in space. Through thorough, rigorous testing and hands-on development, the JSC team were able to address the myriad issues affecting inflatable space structures, including ease of manufacturing, structural integrity, micrometeorite protection, folding, and vacuum deployment. In short, the JSC team had not only proven inflatable structure technology was ready for the space age, but had put the "living" into "living and working in space". For those who, since the project's inception in 1997, had worked on developing inflatable structures, the TransHab project represented a technology that heralded a new era in space. TransHab also captured the imagination of many in the aerospace arena because it promised to change how engineers thought about designing habitats for space. TransHab also broke the volumetric barrier of the exoskeleton spacecraft type by innovating an entirely new, endoskeletal typology and demonstrated the advantages of combining human engineering with aggressive structural innovation and testing at the concept stage. The integrated effort through which this revolutionary habitat was conceived and developed proved its virtue in meeting tremendous challenges by combining innovative design with cutting-edge science that pushed the technological envelope beyond previous design work. With the success of TransHab, there was talk of developing the technology further, integrating sensors, circuitry, and automated components to enable self-deployment of "smart" inflatable structures which could enable a habitat to operate autonomously. Engineers discussed ways to incorporate new breakthroughs in bio-technology to produce a self-healing surface analogous to human skin: fully integrated inflatable "skins" would include sensors to detect, analyze, and repair structural failure on their own. The ground-breaking work by architects and engineers at JSC would lay the technological foundation for structural innovation by many others for years to come. It wasn't to be. TransHab was killed due to budgetary concerns, leaving inflatable habitat technology stalled, until Robert Bigelow bought the technology, which brings us to the Genesis project.

4

Genesis I and *II*

"That's one small step for Bigelow … one giant leap for entrepreneurial space."

Mike Gold, corporate counsel for Bigelow Aerospace, after
witnessing the launch of Genesis I

That lofty pronouncement was made in 2006, when Bigelow Aerospace launched its *Genesis I* habitat into orbit on board a Russian Dnepr rocket, of all things. Not a bad accomplishment for a company that had been founded only seven years previously. But, while Bigelow Aerospace was founded in 1999, it wasn't until 2004 that the company's high-stakes business plan to develop commercial habitats became widely known. With public attention focused on the International Space Station (ISS), Bigelow quietly went about his business developing a mini-Skunk Works for NASA's Johnson Space Center (JSC). At the time, it was hoped that with technical assistance from JSC, Bigelow would be ready to launch his inflatable modules by the end of the decade. But, realizing there was no affordable way of launching his clients to their modules on orbit, Bigelow was spurred to create the US$50 million space launch competition called America's Space Prize in 2004. The goal of the prize was to drive development of affordable manned launch vehicles capable of ferrying up to seven astronauts to low Earth orbit (LEO), and do it by the end of the decade. The winner, apart from winning US$50 million, would be guaranteed first rights on a contract from Bigelow for orbital servicing missions to the company's *Nautilus* modules (later renamed the BA-330). It was anticipated the *Nautilus* would be launched on a Proton-class booster and be ready for launch sometime in 2008.

Nautilus was a key element in Bigelow's *Skywalker* complex (Table 4.1)—a 2005 concept designed to comprise multiple *Nautilus* habitat modules, which would be inflated and connected on orbit. A Multi-Directional Propulsion Module (MDPM) would be fitted to the complex, allowing it to be moved into interplanetary or lunar trajectories. Early assessments of the probability of success of the technology development and challenges of a commercial space station pointed to the importance of a factor largely beyond Bigelow's control and which was to become a bugbear for the company: a transportation system to ferry clients to the station. The transportation issue wasn't helped after the *Columbia*

© Springer International Publishing Switzerland 2015
E. Seedhouse, *Bigelow Aerospace: Colonizing Space One Module at a Time*,
Springer Praxis Books, DOI 10.1007/978-3-319-05197-0_4

Table 4.1. *Skywalker* dimensions.

Crew	Five to seven
Mass	100,000 kilograms
Height	30 meters
Diameter	6.7 meters
Pressurized volume	1,500 cubic meters

accident in 2003, when Bigelow had to compete with NASA for rides on the very pricey Russian Soyuz three-person rocket.

Preceding the *Nautilus* would be two *Genesis* modules, one launched on a SpaceX Falcon 5 and the other on a SS-18 ballistic missile. Once those modules were in orbit, the plan was to launch two *Guardian* modules carrying life-support system demonstration hardware. If all went to plan—something that never happens in the aerospace industry—Bigelow's first customers were expected to be floating around their modules as early as 2010. To that end, Bigelow began courting biotech and pharmaceutical companies in addition to government and civilian users as potential customers. By mid-2004, Bigelow already had various test articles occupying floor space of his facility in North Las Vegas, including three full-scale metallic modules simulating inflatables, a simulator used for testing life-support equipment, two full-sized *Nautilus* inflatable bladders, a 50% scale *Nautilus*, and a one-third scale *Genesis*. Later that year, testing of the *Genesis* development test article began at the Jet Propulsion Laboratory in Pasadena, California, under a Space Act Agreement (SAA) with NASA. Testing included launch load and modal vibration tests, altitude depressurization assessments (to see how the restraint straps held the bladder in a vacuum), and temperature tests. Development of the module went well over the months that followed. The only headache was finding a launch vehicle, the solution to which required some creative out-of-the-box thinking.

In 2004, the plan had been for *Genesis I* to be on board the Falcon 5's maiden flight for a November 2005 launch from Vandenberg Air Force Base, California. Designed to carry 4,500 kilograms into LEO, the Falcon 5 was a five-engine version of SpaceX's first booster, the single-engine, 450-kilogram-payload Falcon 1. At the time, SpaceX was struggling to overcome final major development issues with its Merlin liquid-oxygen/kerosene engine (problems with the flight termination receiver) which required additional expenditure and time. Marketed at US$12 million per launch, the Falcon 5 was a bargain, but the flight never took place and the booster's trouble-plagued development resulted in Bigelow Aerospace searching for another launcher. But, whichever launch vehicle was chosen, Bigelow still required approval from the US government permitting Bigelow Aerospace to launch its inflatable space module. After an extensive eight-month review of the concept, including its construction, materials used, shielding technology, the in-space inflation process, as well as de-orbiting of the test module, in November 2004, the Federal Aviation Administration's (FAA) Associate Administrator for Commercial Space Transportation (AST) gave Bigelow Aerospace approval for flying *Genesis I*. November 2004 proved to be a busy month for Bigelow Aerospace because this was also when "America's Space Prize" (see sidebar) was announced.

America's Space Prize.

Two months after the US$10 million X-Prize had been won by SpaceShipOne, Bigelow Aerospace announced details of the next big space prize—this one for a cool US$50 million. Anyone who wanted to follow in the contrails of Burt Rutan had to build a spacecraft capable of taking a crew of five to an altitude of 400 kilometers and complete two orbits of Earth at that altitude. Then they had to repeat the feat within 60 days. They also had to do this by January 2010. Those were just some of the rules (see below) that governed who won the US$50 million. Other rules required that no more than 20% of the spacecraft's hardware could be expendable. It also had to be able to dock with Bigelow Aerospace's inflatable space habitat and be able to stay on orbit for up to six months. Why the competition? One reason was to try to break the monopoly on crew transport to space held by Russia's Soyuz spacecraft (this was a year following the *Columbia* accident). Another reason was that Bigelow needed a launch company to ferry its customers to their destination and, once the Shuttle stopped flying, that would have meant going head to head with NASA to buy Soyuz spacecraft.

The rules:

- The spacecraft must reach a minimum altitude of 400 kilometers.
- The spacecraft must reach a minimum velocity sufficient to complete two full orbits at altitude before returning to Earth.
- The spacecraft must carry a crew of five.
- The spacecraft must dock or demonstrate its ability to dock with a Bigelow Aerospace inflatable habitat, and be capable of remaining on station for at least six months.
- The spacecraft must perform two consecutive, safe, and successful orbital missions within a period of 60 days.
- No more than 20% of the spacecraft may be composed of expendable hardware.
- The contestant must be domiciled in the US.
- The contestant must have its principal place of business in the US.
- The competitor must not accept or utilize government development funding related to this contest of any kind, nor shall there be any government ownership of the competitor. Using government test facilities shall be permitted.
- The spacecraft must complete its two missions safely and successfully, with all crewmembers aboard for the second qualifying flight, before the competition's deadline of January 10th 2010.

The prize expired in January with no teams making a serious effort to win it. The prize rules that prevented teams from accepting government development funding effectively ruled out the only company that appeared to have a chance at winning: SpaceX.

The FAA–AST approval letter, the first of its kind, was not only a step towards helping nascent space firms move forward, but also paved the way for Bigelow Aerospace to launch *Genesis I* on a private booster—the Falcon 5. But SpaceX's problems led Bigelow to consider other options.

GENESIS I

Between 1992 and 2003, a team of Russian and Ukrainian companies, together with the Russian Ministry of Defense, were developing a commercial space launch system based on the technology of the SS-18 family of intercontinental ballistic missiles (ICBMs) that were being withdrawn from service. The decision to build such a launch system was preceded by various scientific research and design efforts, which resulted in the parties agreeing that the most cost-effective solution for launching payloads into LEO would be a space launch system based on the SS-18 (Figure 4.1) ICBM with minimum modifications of the original missile. The company assigned the role of building and commercially operating this launch system was ISC (International Space Company) Kosmotras and the program was given the name "Dnepr", after the Ukrainian river.

4.1　Dnepr rocket similar to the SS-18 that launched *Genesis I*. Courtesy: Wikimedia

War surplus rockets for sale

The business of converting Soviet-era missiles into cash-paying satellite launches was based on a converted RS-20V "Voyevoda" (Russian for "war chief") known as the SS-18 "Satan" to the Pentagon. In 1999, this ICBM had had more than 20 years of alert duty, when ISC Kosmotras began developing it for more peaceful assignments. To do this, ISC Kosmotras modified the Dnepr-M booster, which was a version of the Dnepr-1 conversion booster, developed from the RS-20 ICBM, NATO designation SS-18 Satan. The first launch of the Dnepr-1 rocket (see sidebar) took place on April 21st, 1999, when a British satellite of the SSTL Company was launched to orbit. More launches followed. For ISC Kosmotras, who planned to use more than 150 SS-18 Satan missiles converted into Dnepr-1 rockets, and for companies searching for low launch costs, it looked like a "win–win" situation. And so it proved. In 2002, two AprizeStar satellites weighing 10 kilograms each were launched as secondary satellites on a Russian Dnepr rocket and two more were launched as secondaries on another Dnepr in June 2004. Then along came Bigelow's *Genesis I*, which was launched at 12:53 UT on July 12th, 2006 into a 95-minute orbit, with an apogee of 561 kilometers, a perigee of 556 kilometers, and inclination of 64.5°. The 1,300-kilogram craft successfully inflated about two hours after launch to its normal cylindrical size of 2.4 meters by 4.5 meters.

Dnepr Launch Vehicle Specifications

The Dnepr launch vehicle is based on the SS-18 liquid-fuelled ICBM and has a three-stage plus Space Head Module (Figure 4.2) in-line configuration. The launch vehicle's first and second stages are original SS-18 stages and used without any modification. The Dnepr third stage is an original SS-18 third stage with an upgraded control system that enables implementation of the required flight program of the first, second, and third stages, forming and issuing commands to payload and Space Head Module separation devices and getting the third stage and remaining Space Head Module elements off the injection orbit after payload separation. Before the Dnepr entered commercial service, it was used by the Strategic Rocket Forces which launched the ICBM version over 160 times with 97% reliability. The rocket holds the record for the most satellites orbited in a single launch when it placed 32 satellites and an experiment package bolted to the upper stage into LEO on November 21st, 2013.

Specifications of the Dnepr launch vehicle:

- Weight at lift-off: 211 tonnes
- Propellant: amyl + heptyl
- Number of stages: three
- LV diameter: 3 meters

(continued)

(continued)

- LV length: 34 meters
- Injection accuracy:

  ◦ for orbit altitude: ± 4.0 kilometers
  ◦ for orbit inclination: ± 0.04°
  ◦ for right ascension of the ascending node: ± 0.05°
  ◦ for orbit inclinations: 64.5; 98.0°
  ◦ for flight reliability: 0.97

- Operational environments:

  ◦ longitudinal G-load: up to 7.5
  ◦ lateral G-load: up to 0.8
  ◦ integral level of sound pressure: up to 140 dB

Mike Gold

At this point in the *Genesis* story, it is necessary to introduce Mike Gold, who was mentioned briefly at the end of the preceding chapter, for it was he who, more than anyone, was responsible for realizing the launch of a commercial module on a refitted nuclear ICBM. Gold's space pedigree extends back to 1996 when he became a law clerk for NASA's Langley Research Center. It was a position that confirmed he could combine his interest in space with his legal work. He went on to represent space start-up companies and worked on spaceport initiatives before being retained by Bigelow Aerospace as corporate counsel, which initially involved the writing of a White Paper dealing with the development of space and how NASA was competing with the private sector. The White Paper was a catalyst for discussion between Bigelow and NASA that eventually led to the signing of a memorandum of understanding between the agency and the Las Vegas start-up. Today, Gold wears many hats, including corporate counsel, handling many of the company's legal issues, dealing with the company's interaction with NASA, as well as with other federal entities such as the FAA. He also oversees many of Bigelow Aerospace's interactions with other corporations such as SpaceX, Kosmotras, and Lockheed Martin.

One of Gold's first tasks as Bigelow's corporate counsel was to navigate the FAA payload approval process for *Genesis I*. This was to prove problematic because Bigelow chose to use a Russian launch provider. This decision brought the *Genesis I* project under the umbrella of the State Department's export regulations known as the International Traffic in Arms Regulations, or ITAR. Gold was shocked by the labyrinthine bureaucracy involved in just getting *Genesis I* to Russia—a process that took the best part of a year. After wading through a paper blizzard of administrative rules and procedures, Gold finally procured a Technical Assistance Agreement (TAA), allowing Bigelow Aerospace to engage with foreign entities, provided monitors from the Defense Technology Security Administration (DTSA) were present. With DTSA officials (who were paid by Bigelow at an hourly rate of US$130) watching his every move, Gold had an uphill task ahead of him to handle the

4.2   Space head module. Courtesy: Wikimedia

legal and contract negotiations with the Russians. The problem was ITAR or, more specifically, the failure of ITAR to distinguish between benign, commercially available technologies and those with military applications. Let's use the coffee table/test stand mentioned in Chapter 1 as an example here. The *Genesis I* test stand was a round piece of metal with four legs sticking up. Flip it over, cover it with a tablecloth, and you'd have a coffee table. But, as part of the TAA, Gold had to have two security guards standing watch over the table in Russia … 24/7! Why? Because of the security risks of coffee table technology falling into the wrong hands. Such are the bureaucratic eccentricities of the export-control process![1] It was just one of many challenges Gold faced during his time in Russia, where he reckons to have spent the best part of three years handling the legal and contract negotiations and acting in the role of launch campaign manager on the ground. Living in a converted barracks built at the Yasny launch base, Gold overcame the difficulties until finally the day came for the spacecraft to be moved from the base to the silo ready for launch. The plan had been for *Genesis I* to be transported on board a missile carrier with armed guards, but the Russians amended the timetable without informing Gold, probably because they didn't want an American hitching a ride on a missile transporter—a Russian implementation of their version of ITAR. The problem was that a spring melt was underway, which had deepened potholes in a road already made treacherous by mud and water. Gold became concerned because he thought *Genesis I* might be damaged en route, so he jumped in front of the convoy and immediately realized it was a bad idea when he heard the clicks of guns being made ready. Fortunately, his Kosmotras representatives rescued him, and put Gold and his team in a van in front of the transporter. After that incident, things went a lot smoother and *Genesis I* launched successfully.

Some media heralded the successful launch of the experimental *Genesis I* module as another step in the ongoing evolution of the space tourism industry, although space tourism is not, and never has been, part of Bigelow's business plan. Other media gave the Russians credit for turning what was once their most feared ICBM into a moderate-to-low-cost space launch vehicle, noting that *Genesis I* was delivered into orbit with an accuracy of about 400 meters, which suggested the US estimate's of the SS-18's accuracy hadn't been far off the mark. But perhaps the most important aspect of the *Genesis I* mission was not just demonstrating that future commercial astronauts would have a place to stay when they got up there, but also showing investors that the commercial spaceflight industry had the potential to become a business. The *Genesis I* design was also important in attracting

---

[1] The problem Gold encountered when dealing with DTSA and ITAR was that the system didn't encourage officials to make commonsense judgments when dealing with off-the-shelf technologies. In short, there was very little discretion built into the system, which would probably work better as a bifurcated process, so that commercially available technology—a coffee table!—is not treated in the same way as militarily sensitive hardware. Such a system would distinguish one from the other and, in so doing, allow the DTSA to focus more of their resources on technologies that legitimately need watching. The way the system is regulated today has drawn criticism from many US companies and academia, who argue ITAR regulatory muscle impedes US trade and hinders science exchange. On the flip side, the Department of State contends ITAR is a necessity, with the various rules and regulations imposed having limited impact.

sovereign nations interested in flying their own astronauts, many of whom would be conducting science. To that end, Bigelow engineers designed *Genesis I* to be as space-worthy as possible, installing a window and instruments that included dosimeters, microphones, and interior cameras.

Genesis I on orbit

Shortly after the launch, Bigelow Aerospace's Mission Control Center in North Las Vegas confirmed telemetry from the spacecraft was being received by ground controllers. Incidentally, unlike NASA, Bigelow Aerospace does not track its spacecraft itself, relying instead on the US Department of Defense, which observes and measures the orbits of thousands of near-Earth objects and publishes the results. Bigelow Aerospace crunches these measurements using off-the-shelf orbital-prediction software to predict the 5–12-minute windows when *Genesis I* (it's still up there) is within line of sight of the company's ground stations in Las Vegas and Arlington, VA. Only during these windows can the control room receive telemetry and transmit commands. In the eyes of many in the commercial space industry, the launch made Bigelow Aerospace the odds-on favorite to create a low-cost, LEO space complex accessible to the commercial sector. After attaining orbit, computer-controlled air-pressure tanks activated and expanded the pre-folded structure into its watermelon shape—a process that took 15 minutes. Even as *Genesis I* circled Earth, work was underway for a *Genesis II* launch. That flight was expected to lead to lofting a larger, more sophisticated module called *Galaxy*. But let's take a closer look at *Genesis I* (Table 4.2).

Systems

As a stand-alone orbiting capsule, *Genesis I* was an entirely new class of spacecraft, with a new complement of power, control, communication, and other flight systems (Table 4.3), many of which were adapted from existing aerospace systems. Given the highly competitive environment of commercial space development, Bigelow Aerospace has been short on details, although it is known that, to enhance reliability, engineers made extensive use of a technique called *dissimilar redundancy*. This meant *Genesis I* was equipped with two means of performing critical functions, such as monitoring internal conditions or distributing electrical power from its solar panels. Basically, instead of having back-ups, *Genesis I* had physically independent systems, not only making it a robust spacecraft, but also allowing easy evaluation of Product A versus Product B. In fact, as many as a third of the systems aboard *Genesis I* were there purely for evaluation for use on future flights. For example, *Genesis I* had two communications systems, with nearly identical installations at each end of the spacecraft, which meant, no matter which end of the spacecraft was pointing towards Earth, there was always full communications capability. This redundancy was needed because *Genesis I* didn't have a way to control its attitude, relying on gravity to torque it into position with its long axis pointed towards the ground.

In addition to having the benefit of dissimilar redundancy, *Genesis I* engineers also had the luxury of not having to worry about weight constraints, the bane of all aerospace engineers, who usually have to fret about shaving each and every nonessential milligram from

Table 4.2. *Genesis I* launch specifications.

Launch date	July 12th, 2006, 14:53:30 UTC
Launch location	ISC Kosmotras Space and Missile Complex, Russia
Length	4.4 meters
Diameter	2.54 meters
Usable volume	11.5 cubic meters
Solar arrays	Eight
Bus	• 2 × Battery charge regulators • 4 × Two-axis digital Sun sensors • Analog-to-digital telemetry board • Custom 28-volt NiCad battery • Integrated flight computer • RAM board • 2 × Three-axis magnetometers • 3 × Micro torque rods • 6 × General purpose terminals • 2 × Custom vision systems • 13 × Ethernet and Firewire digital cameras • 120 Triple-junction, GaAs solar panels
Shell skin	Multilayer system, 15 centimeters thick
NORAD identifier	29252
Earth orbits	Once every 96 minutes
Speed	27,242 kilometers per hour (7.59 kilometers per second)
COSPAR ID	2006-029A
Re-entry	2013–2019
Mass	1,360 kilograms
Atmospheric pressure	51.7 kilopascals
Perigee/apogee	531/549 kilometers
Orbital inclination	64.51°
Orbital period	95.44 minutes
Orbits per day	15.08

Table 4.3. *Genesis I* systems.

Solar arrays	Eight
Bus	• 2 × Battery charge regulators • 4 × Two-axis digital Sun sensors • Analog-to-digital telemetry board • Custom 28-volt NiCad battery • Integrated flight computer • RAM board • 2 × Three-axis magnetometer • 3 × Micro torque rods • 6 × General purpose terminals • 2 × Custom vision systems • 13 × Ethernet and Firewire digital cameras • 120 Triple-junction, GaAs solar panels
Shell skin	Multilayer system, 15 centimeters thick
Mass	1,360 kilograms
Atmospheric pressure	51.7 kilopascals

their designs to meet exacting launch limits. But, because Bigelow Aerospace had chosen to launch *Genesis I* on the Dnepr rocket, weight-saving wasn't a factor because the launch weight of about 2,000 kilograms used only half of the Dnepr's capability. That said, even with the generous weight margin, some features had to be dropped. For example, Bigelow had planned to put his wife's name (Diane) on the outside of the module in glowing lights but, for technical reasons, engineers had to take it off. Instead, the Russians wrote her name on the fairing of the rocket, next to the American flag.

Genesis I is outfitted with eight solar panel arrays, four on each end of the craft, which produce one kilowatt of power and maintain a 26-volt battery charge. The spacecraft carries 13 cameras, seven externally to monitor the physical condition of the spacecraft, such as the outer shell and solar arrays, and six internally to photograph objects and experiments. Atmospheric pressure is 51.7 kilopascals and a passive thermal control system maintains an average temperature of 26°C. The module also features a Micro Meteorite Orbital Debris (MMOD) shield comprising multilayer insulation with interstitial foam to provide loft, a load-bearing restraint layer, and an air barrier folded in launch configuration and restrained with straps. The deployment system comprised retention straps released using pyrotechnic cutters controlled by an on-board flight computer. Inflation was achieved via a single tank fitted with redundant solenoid valves.

All data recorded on board the spacecraft are downloaded as encrypted files during passes over ground sites. The data sampling rate can be scaled up and down as necessary to prevent build-up of data. At each end of the spacecraft are redundant omni-directional antennas: UHF/VHF for duplex command and telemetry, and S-band for photo downlink. The module's attitude control subsystem (ACS) ensures the spacecraft is pointing in the right direction, although the mission design requires minimal pointing requirements. As *Genesis I* orbits Earth, it rotates about its long axis ensuring a benign thermal environment—a so-called "rotisserie" effect. The attitude of the spacecraft is sensed with magnetometers and four Sun sensors while actuation is provided with magnetic torque rods: two are mounted in the longitudinal axis and one each is mounted in the y-axis and the z-axis.

The spacecraft's avionics subsystem features a controller area network (CAN) bus (Table 4.2) and nearly every component of the bus is CAN-compliant. With each component having at least one on-board processor, the spacecraft is a network of small computers, each looking after itself and its functions: from the battery control regulators and general purpose terminals used for collecting telemetry to individual Sun sensors, every CAN-compliant device is networked together by interface cables which route power and data between nodes.

Genesis I payload

Genesis I carries an assortment of payloads, including photographs, toys, and cards from Bigelow employees and a life-sciences payload of Madagascan hissing cockroaches and Mexican jumping beans together with NASA's GeneBox. A shoebox-sized payload - GeneBox - contains a miniature laboratory designed to analyze how weightlessness affects

Table 4.4. GeneBox specifications.

Weight	4.6 kilograms
	1.4-kilogram mounting bracket
	3.2-kilogram GeneBox and power/data interface box
Dimensions	38 × 10 × 10 centimeters
Power consumption	2 watts (4 watts peak)
Control	Two PIC 18F6720 processors
Mounting	Three clamps around longeron
Data collection	• System times
	• Control set points
	• Status indicators
	• Temperatures
	• Payload pressure
	• Currents
	• Voltages
	• Relative humidity
	• Radiation
	• Vibration
	• Fluorescence
	• Optical density data

genes in microscopic cells and other small life forms. The micro-laboratory also includes sensors and optical systems that can detect proteins and genetic activity. The payload, which NASA estimated cost US$1 million to fabricate, test, *and* prepare for flight, was integrated and launched by Bigelow Aerospace at no cost to NASA. GeneBox (Table 4.4) was an adaptation of the GeneSat free flyer nanosatellite, which meant that, to take advantage of Bigelow Aerospace's offer to be accommodated within *Genesis 1*, the GeneSat free flyer design had to be reconfigured. This meant removing GeneSat's solar cells and batteries, and adding a GeneBox-to-*Genesis I* power/data interface box.

Mission status

Much of the spacecraft's time on orbit has been uneventful, with the exception of suffering a radiation event in December 2006 as a result of a solar storm. Fortunately, Mission Controllers were able to restart the system in time, though the situation was just one fault away from the spacecraft being declared dead. Despite this, no lasting damage occurred and the spacecraft was soon back to operating healthily. On May 8th, 2008, 660 days after launch, *Genesis I* completed its 10,000th orbit, by which time it had logged more than 430 million kilometers—the equivalent of 1,154 return trips to the Moon. The spacecraft had also taken more than 14,000 images, including images of all seven continents. Although the design life of the avionics was only six months, the avionics systems worked for more than two and a half years before failure and, in February 2011, Bigelow Aerospace reported the vehicle had "performed flawlessly in terms of pressure maintenance and thermal control-environmental containment". The orbital life is estimated to be 12 years, with a gradually decaying orbit resulting in re-entry into Earth's atmosphere and burn-up expected. But, in 2014, *Genesis I* was still operating nominally, dutifully transmitting data about its temperature, hull integrity, power levels, and overall health. It was the spacecraft's seventh year on orbit—a clear demonstration of the long-term viability of expandable habitat technology in an orbital environment.

GENESIS II

After *Genesis I*, Bigelow Aerospace had plans to launch the next in the company's series of prototypes: *Genesis II*. *Genesis II* was to be followed by the launch of test modules twice a year until the human-rated version was perfected. At the time of the *Genesis II* launch, the company envisaged having up to five active vehicles in space, each with a life expectancy of three to seven years. By collecting and analyzing data about the robustness of the vehicle's systems and the integrity of the hull, and with the ability to rapidly deploy and test improvements in space, Bigelow Aerospace believed it could orbit a habitable module within three years.

The months leading up to the *Genesis II* launch were almost as much of an administrative endurance test as the process experienced prior to the *Genesis I* launch. The transporting of *Genesis II* to Russia for launch was the end result of seemingly endless months of regulatory processes due to restrictions imposed by ITAR. After leaving North Las Vegas, *Genesis II* made a stopover in Luxembourg before being flown on an Antonov An-124 to Orsk, Russia, and transported over ground to the Dombarovskiy base. *Genesis II* was finally moved into the Assembly, Integration, and Test Building on March 29th, 2007 (the planned launch date had been August 6th, 2006), but launch proved a frustrating experience, ISC Kosmotras delaying the launch four times (April 1st, April 19th, April 26th, and May 23rd) due to technical and scheduling concerns. Finally, almost exactly a year after the *Genesis I* launch, *Genesis II* was launched (Table 4.5) from the same site atop a converted RS-20V Voyevoda (SS-18 Satan) rocket on June 28th, 2007. This time, the launcher placed its payload within 100 meters of its planned parking orbit.

Genesis II on orbit

Although externally, *Genesis II* looked like a twin of *Genesis I*, *Genesis II* contained several systems not flown on its predecessor, such as additional cameras, sensors, a Biobox, and a reaction wheel. *Genesis II* also allowed the general public to participate in the mission through the company's Fly Your Stuff program, of which more later. *Genesis II* also sported more cameras than *Genesis I* (22—nine more than *Genesis I*), including articulated cameras with dual FireWire and Ethernet interfaces, as well as a wireless boom camera for exterior shots. Space-to-ground communications are provided by UHF, VHF, and S-band antennas, and magnetic torque rods, GPS, Sun sensors, and a reaction-wheel system provide attitude control and stabilization. On December 12th, 2007, Bigelow Aerospace confirmed *Genesis II* was in good health. The cameras had been tested and more than 4,000 images had been taken. The spacecraft's attitude control systems and solar arrays were operational and internal pressure was holding between 69.6 and 72.4 kilopascals,

Table 4.5. *Genesis II* launch specifications.

Launch date	June 28th, 2007, 15:02:00 UTC
Launch location	ISC Kosmotras Space and Missile Complex, Russia
Length	4.4 meters
Diameter	2.54 meters
Usable volume	11.5 cubic meters
Solar arrays	Eight
Bus	• 2 × Battery charge regulators
	• 4 × Two-axis digital Sun sensors
	• Analog-to-digital telemetry board
	• Custom 28-volt NiCad battery
	• Integrated flight computer
	• RAM board
	• 2 × Three-axis magnetometer
	• 3 × Micro torque rods
	• 6 × General purpose terminals
	• 2 × Custom vision systems
	• 13 × Ethernet and Firewire digital cameras
	• 120 Triple-junction, GaAs solar panels
Shell skin	Multilayer system, 15 centimeters thick
NORAD identifier	31789
Earth orbits	Once every 96 minutes
Speed	27,242 kilometers per hour (7.59 kilometers per second)
COSPAR ID	2007-028A
Re-entry	2013–2019
Mass	1,360 kilograms
Atmospheric pressure	69.6–72.4 kilopascals
Perigee/apogee	520/561 kilometers
Orbital inclination	64.51°
Orbital period	95.44 minutes
Orbits per day	15.07

the variation caused by *Genesis II* moving in and out of sunlight as it orbited. On April 23rd, 2009, after *Genesis II* had been in space for 665 days, Bigelow Aerospace announced the habitat had completed 10,000 orbits, and traveled over 434 million kilometers. *Genesis II*'s orbital life is estimated to be 12 years, with a gradually decaying orbit resulting in re-entry into Earth's atmosphere.

Payload

Genesis II was the first Bigelow Aerospace module to carry personal items under the company's "Fly Your Stuff" campaign—an initiative where, as the name suggests, people could fly their photos or small items for US$300. Fly Your Stuff was not only an exercise in media outreach to gauge whether the public was interested in what Bigelow was doing, but also an effort to diversify the company's revenue streams. From a business standpoint,

Fly Your Stuff made sense because, with the vagaries of funding expensive launches, no company can depend on one income stream: by testing the waters with this initiative, Bigelow was gaining an understanding of what income streams could be generated. For those interested in entertainment, *Genesis II* carries a Space Bingo game (Bigelow is based in Las Vegas after all), a module containing a set of Bingo balls randomly manipulated by a system of fans and levers, resetting after 40 balls had entered play.

Genesis II's science payload is a life-sciences module termed BioBox that includes habitats for three organisms: the Madagascar hissing cockroach; the South African flat rock scorpion, *Hadogenes troglodytes*; and a colony of seed-harvester ants, *Pogonomyrmex californicus*, along with the queen ant for colonization purposes. This system includes automated food and water delivery systems, and fans keep fresh air available by circulating internal air with that inside the rest of the spacecraft.

Systems

Genesis II features a number of improvements over the first module. In addition to the standard guidance control systems used on its predecessor, *Genesis II* has reaction-wheel assemblies and a precision measurement system, used to affect the spacecraft's rotation rate and angular momentum without expending fuel. Instead of the single-tank inflation system used on the first craft, *Genesis II* employed multiple tanks for added reliability and to allow for more finely tuned gas control. Additional layers were added to the outer shield for increased protection and thermal management. Finally, the on-board sensor suite was enhanced with additional sensors for pressure, temperature, attitude control, and radiation detection, to help determine the impact of the orbital environment on the integrity of shipboard systems.

TOWARDS LARGER MODULES

At the time of the *Genesis II* launch, Bigelow Aerospace planned to launch *Galaxy* in 2008, another pathfinder module that built on the *Genesis* vehicles, before flying *Sundancer*, its first crew-rated spacecraft, in 2010.

Galaxy

Galaxy was to have 45% more habitable space than the *Genesis* module, with a pressurized volume of 16.7 cubic meters. The module was intended to be an evolutionary step between the *Genesis*-class modules and the standard, human-habitable complex modules. It was also to provide first-flight experience for technologies being developed for future commercial space complexes. Some of the technology advances that were to be implemented on *Galaxy* included:

- advanced on-board avionics such as increased performance, decentralized processing, and increased redundancy
- Environmental Control Life-Support System (ECLSS); *Galaxy* was not designed to support astronauts, but to serve as a test bed to flight-qualify critical elements of crew-support systems

- upgraded attitude determination and control system; this was to be an upgraded version of the *Genesis*-class guidance, navigation, and control system to provide greater torque-control authority and momentum management with precision attitude sensing
- increased communications bandwidth (capability of a real-time video downlink)
- a more robust air barrier
- power system improvements that were to include larger and more efficient articulated solar arrays and high-performance battery technologies
- structural upgrades were to be made to the primary structure and the software, which would support scalability to the larger spacecraft; the addition of an access hatch and an additional, larger viewing port was also planned.

Galaxy never flew. In August 2007, Bigelow Aerospace announced that, due to rising launch costs (three times more expensive than for previous launches) and the successful *Genesis* missions, the *Galaxy* spacecraft would not be launched. Instead, the plan was to advance *Sundancer's* schedule.

Sundancer

The three-person, 180-cubic-meter-volume *Sundancer* was expected to be fitted with a connecting node and propulsion bus to lay the foundation to support Bigelow Aerospace's even larger BA-330 module. The BA-330 was expected to have a 330-cubic-meter habitable volume and the plan was to dock with *Sundancer* and its node-propulsion bus by 2012. Basically, the module represented a step-by-step increase in size—a strategy designed to not only establish the technology, but also help build the business case for orbiting inflatable modules. Such an incremental approach made sense because, until the *Genesis I* and *II* flights, no expandable system had ever been tested in an orbital environment: by flying several modules, Bigelow hoped to demonstrate and validate the core technology, such as the inflation process, as well as the durability and longevity of the vehicle when exposed to the orbital environment for a period of years.

Sundancer (Figure 4.3) was expected to have a mass of about 8,200 kilograms and would have been equipped with life-support systems, attitude control, three windows, on-orbit maneuverability, re-boost, and de-orbit capability. The module was planned to be Bigelow's first manned test bed to assess systems to be used in the company's commercial space station efforts. It was also planned to form the first piece of the first commercial space station. Plans were to launch and dock a combined propulsion bus and central node to *Sundancer* in 2014. By 2010, Bigelow was show-casing full-sized space station mockups (Figure 4.4) sitting on its warehouse floor and the plan was to launch and assemble the first private space station by 2014. Paying customers would arrive a year later. Then, in 2016, a second, larger station would follow. The two Bigelow stations would be home to 36 people, or six times as many as who live on the ISS. The business plan had Bigelow buying 15–20 rocket launches in 2017 and in each year thereafter, providing ample business for companies like SpaceX. A new market would open and access to space would catalyze the potential of American industry. And the best part about this future was that Bigelow was the only game in town.

4.3 *Sundancer* module. Courtesy: Boeing

4.4   Robert Bigelow and NASA Deputy Administrator, Lori Garver, in front of a Bigelow module. Courtesy: Bill Ingalls/NASA

"Similar to the beginnings of commercial telecommunications or aviation, we are witnessing the dawn of a new industry that will change the way we live life here on Earth. Robust and reliable access to microgravity will impact fields as diverse as pharmaceutical development to fuel production, representing a broad and substantial technological leap forward. The countries and companies that have the foresight to develop early access to orbital capabilities and infrastructure will become the economic and technological leaders of the future."

Robert Bigelow

Space Station Alpha

The space station mock-up displayed on the factory floor was first referred to as Space Complex Alpha in 2010 (Table 4.6), which initially comprised two *Sundancer* modules and one BA-330 module, a configuration that later evolved into two BA-330 modules. In October 2010, Bigelow Aerospace announced it had agreements with six sovereign nations to utilize the on-orbit facilities of the commercial space: the UK, The Netherlands, Australia, Singapore, Japan, and Sweden. In 2011, the United Arab Emirate of Dubai (UAE) joined

Table 4.6. Building a space station in seven steps.

1	Unit 1: First *Sundancer* module (pressurized volume of 180 cubic meters) launched early 2014 unmanned
2	Unit 2: Commercial crew capsule arrives with Bigelow Aerospace astronauts to deploy *Sundancer* and carry additional supplies
3	Unit 3: Supplemental power bus and docking node arrives
4	Unit 4: Second *Sundancer* module launched
5	Unit 5: Second commercial crew capsule brings additional crew and supplies, and provides a lifeboat capability for crew return to Earth
6	Unit 6: BA-330 (pressurized volume of 330 cubic meters) launched
7	Unit 7: Third commercial crew capsule brings additional supplies and provides a double-redundant, robust solution for astronaut re-entry

the list (see sidebar). To service the anticipated clients, Bigelow Aerospace began building a large production facility in North Las Vegas, to produce space modules. The plan at the time was to use three production lines to transition the focus from research and development to production. In 2010, Bigelow expected to hire more than 1,000 employees to staff the plant, and planned to start production in early 2012. It wasn't to be: in October 2011, Reuters reported Bigelow had reduced its 115-member workforce to just 51 because of delays in developing commercial vehicles needed to fly people to orbit (a topic discussed in Chapter 6). But 2011 wasn't all bad news because, earlier in the year, the International Space Station Program (ISSP) managers at NASA/JSC held a two-day technical meeting to discuss the prospect of adding a Bigelow Aerospace inflatable module to the ISS (this proposal had been originally outlined at NASA's Exploration Enterprise Workshop, held in Galveston, Texas, in May 2010). The module destined for the ISS, like pretty much everything that is flown to the orbiting outpost, had its own acronym: BEAM.

On January 31st, 2011, the Emirates Institution for Advanced Science and Technology (EIAST) and Bigelow Aerospace signed a Memorandum of Understanding (MoU) to explore joint efforts to establish a commercial human spaceflight program for Dubai and the UAE. On signing the agreement, Mr. Al Mansoori, Director General of EIAST, said "The partnership of EIAST with Bigelow Aerospace is a critical next step forward for the organization in exploring the potential for human spaceflight programs. The MoU will not only elevate Dubai to a stronger global platform as a facilitator of commercial human spaceflight but also create more opportunities for people anywhere in the world to take advantage of our initiatives to experience the marvels of space travel." Established in 2006, the primary roles of the EIAST are to promote the culture of advanced scientific research and technology innovation in Dubai and the UAE, create an internationally competitive base for human skills development, position Dubai and the UAE as a science and technology development hub among advanced nations, and establish international collaborative links with industry and research organizations. EIAST focuses on four scientific programs: Space; Astronomy; Energy; and Environment and Water Research.

5

Bigelow Expandable Activity Module

"Today we're demonstrating progress on a technology that will advance important long-duration human spaceflight goals. NASA's partnership with Bigelow opens a new chapter in our continuing work to bring the innovation of industry to space, heralding cutting-edge technology that can allow humans to thrive in space safely and affordably."

Lori Garver, NASA's Deputy Administrator, announcing plans to test inflatable space habitat technology on the International Space Station

For all their success, the *Genesis* missions were but a prelude to Bigelow's BEAM (Bigelow Expandable Activity Module) and BA-330 behemoth. The BA-330, so numerically named because it will boast a habitable volume of 330 cubic meters, represents a 210% habitability increase over the International Space Station's (ISS's) Destiny laboratory. The BA-330 has the capability to support scientific and industrial microgravity research and, despite its cavernous interior, its textile and Vectran construction will push its weight no higher than 22,000 kilograms. Solar arrays and batteries will provide electrical power, with avionics to support navigation, orbital re-boost, docking, and maneuvering, and an environmental control system to sustain up to six human occupants. But, before the BA-330, a smaller inflatable must be tested: the BEAM. The announcement, on January 16th, 2013, went on to describe how NASA had awarded Bigelow Aerospace a US$17.8 million contract to demonstrate its BEAM technology on a mission slated to fly to the orbiting outpost in 2015, where it will remain for two years. BEAM will be launched to the ISS on SpaceX's Dragon spacecraft (Figure 5.1). The module will be part of the cargo delivered to the station during the eighth resupply flight SpaceX will conduct under the Cargo Resupply Services contract the company has with NASA. If all goes well, astronauts on board the ISS will use the station's robotic arm to install BEAM to the aft port of the Tranquility node (Figure 5.2). Once attached to Tranquility, BEAM will be activated and the module will inflate via a pressurization system that is part of the module.

BEAM will remain attached for two years, during which time it will be tested by controllers on the ground and the crew on board the ISS. Engineers will be particularly interested in measuring the module's leak rate, so BEAM will be fitted with an assortment of instruments to monitor its performance. In addition to data generated by on-board instruments, ISS crewmembers will enter the module periodically to see how it is performing. Then, after two years, BEAM will be jettisoned from the station and burn up as it re-enters Earth's atmosphere.

© Springer International Publishing Switzerland 2015
E. Seedhouse, *Bigelow Aerospace: Colonizing Space One Module at a Time*,
Springer Praxis Books, DOI 10.1007/978-3-319-05197-0_5

5.1 SpaceX's Dragon capsule may be one of the means by which Bigelow's clients travel to their orbiting habitats. Courtesy: SpaceX

5.2 Artist's rendering of the BEAM attached at the International Space Station (ISS). The BEAM will likely be installed onto the Node 3 module of the ISS due to the limited number of berthing ports available on the ISS for expansion. This is due to the requirements for two open ports on the ISS for cargo resupply vehicles and also two NASA ports for commercial crew vehicles. This leaves only Node 3, since the Node 3 Forward and Zenith ports are unusable due to tight clearances with the P1 and Z1 Trusses. The Node 3 Aft location will afford the BEAM some additional micrometeorite protection by "trailing" it behind Node 3 and the Japanese Pressurised Module, which will bear the brunt of any debris. Courtesy: Bill Ingalls/NASA

ADVANCED EXPLORATION SYSTEMS

BEAM falls under NASA's Advanced Exploration Systems (AES) program dedicated to developing new technologies aimed at future crewed missions, and NASA hopes BEAM will help the agency's human spaceflight ambitions for destinations beyond low Earth orbit (LEO).

The AES program, which is uniquely related to crew safety and mission operations in deep space, consists of about 20 small projects that target high-priority capabilities needed for human exploration such as advanced life support, deep space habitation, crew mobility, and extravehicular activity (EVA) systems. The rationale driving the AES program is that early integration and testing of prototype systems, such as BEAM, will reduce risk and improve affordability of exploration mission elements. To realize that goal, prototype systems developed within the AES program will be demonstrated in ground-based test beds, field tests, underwater tests, and flight experiments on the ISS. For example, one of the programs within AES is the Habitation Systems Project (HSP), whose charter is to define and mature a Deep Space Habitat (DSH) that will enable human exploration to multiple destinations. HSP is a project involving a multi-center team of NASA architects, scientists, and engineers, all working together to develop sustainable living quarters, workspaces, and laboratories for next-generation space missions. The project team already have a DSH mock-up (Figure 5.3) installed inside Building 4649 at Marshall Space Flight Center.

5.3 Deep Space Habitat mock-up. Courtesy: NASA

After developing and field testing a DSH concept demonstrator—the Habitat Demonstration Unit—the team is constructing an ISS-derived DSH Concept Demonstrator, which is being constructed with mock-ups of modules currently in use on the ISS. The demonstrator entails outfitting a full-sized module with operational systems to allow the evaluation of living and working in the habitat. For the notional concept of an ISS-derived DSH, the following elements are included:

- an ISS Lab-sized element and a Multi-Purpose Logistics Module (MPLM) to provide pressurized volume for working and living space;
- a Utility Tunnel including an airlock to traverse between elements and allow EVAs;
- a Multi-Mission Space Exploration Vehicle (MMSEV).

INFLATABLE SPACESHIPS

The last item on the list brings us to Bigelow's BEAMs and takes us back to 2010 when Mark Holderman and Edward Henderson, two brilliant and visionary engineers of NASA's Technology Applications Assessment Team (TAAT), designed the NAUTILUS-X, MMSEV. The rather cumbersome (even by NASA standards) acronym stands for Non-Atmospheric Universal Transport Intended for Lengthy United States–Xploration, Multi-Mission Space Exploration Vehicle. The NAUTILUS-X (Figure 5.4) isn't the sexiest spacecraft ever created but, despite its gangly appearance, the concept is functional. Designed to support a crew of six in deep space for up to two years, the spacecraft features an integrated (inflatable) centrifuge and the habitation elements are Bigelow-style expandable modules similar to BEAMs. All in all, the NAUTILUS-X constitutes robust outside-the-box thinking that represents by far the most exciting spacecraft NASA has designed in decades. Let's take a tour.

Going from aft to bow, we see the distinctive propulsion system ringed by fuel tanks and radiators. Forward of the propulsion system are the expandable modules, containing logistical stores and hangars containing the descent vehicle. Above the propulsion system, you can see the primary communications dish and forward of that are the EVA-pods and science probe craft. The environmental closed life-support system (ECLSS) is the vertical expandable module perched above the forward logistical store. Forward of this is the distinctive shape of the centrifuge and the photovoltaic arrays. Finally, at the sharp end of this visionary spacecraft are the command and observation deck, primary docking port, and the adaptable full-span remote manipulator system. Those with an engineering background may notice the lack of a thermal rejection capability. While the NAUTILUS-X is being constructed, the thermal load/rejection and management issue will be resolved by the tried and tested classic basting roll maneuver but, for its deep space trek, a different and yet-to-be-determined thermal management solution will probably be implemented. Engineers will also notice the absence of an exo-truss, necessary for managing and transmitting the load path from the propulsion system. That is because the issue of load path negotiation has yet to be resolved in this design. While science-fiction aficionados may take exception to the NAUTILUS-X's bulky

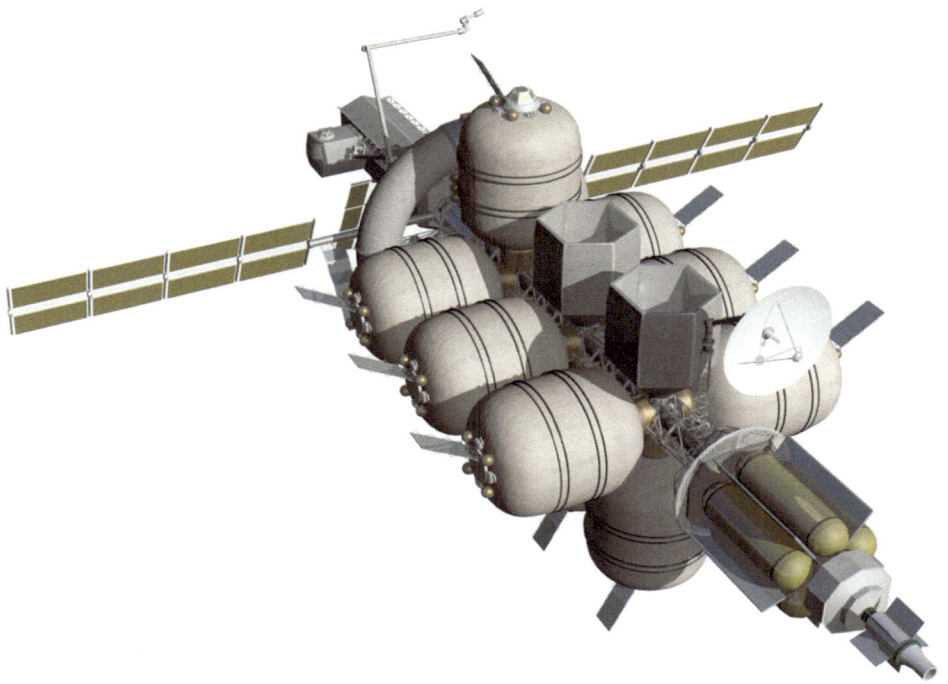

5.4   One application of the inflatable habitat concept: the NAUTILUS-X, MMSEV. Courtesy:
Mark Holderman/NASA

design, you need to remember that many of its design factors are constrained by the
modular method of the vehicle's construction. Also, it is worth pointing out that, while
many spacecraft in science-fiction films may look elegant, there is often little regard to
sound engineering principles. But back to the inflatable elements.

An inflatable centrifuge? Absolutely. While the ISS could be expanded considerably
using BEAMs and/or BA-330s, most astronauts would probably prefer a partial gravity
capability. Those of you who have followed the history of the ISS will remember that the
original design featured a centrifuge accommodations module (CAM) (see sidebar)—an
element presently gathering dust in a car park somewhere in Japan. While an inflatable
centrifuge was designed for the NAUTILUS-X, the engineers realized they needed to
assess and characterize the influences and effects of the centrifuge relative to human reactions, mechanical dynamic responses, and influences. The best place to do that? The ISS,
naturally. In common with all inflatable modules, the centrifuge (Figure 5.5) would be a
low-mass structure, featuring a rotating hub/transit tunnel. It could be launched on a Delta
IV or Atlas V launcher for a cost of between US$83 and US$143 million.

5.5   Inflatable centrifuge. Courtesy: Mark Holderman/NASA

CAM

The CAM flight disappeared from NASA's flight manifest in spring 2001, when the US$4.8 billion ISS overrun exploded in the face of the Bush administration, which responded by scaling back the ISS. The CAM flight was restored to the flight manifest after incoming administrator Sean O'Keefe straightened out the ISS finances. After the 2003 *Columbia* accident, NASA released a 28-flight manifest that completed ISS assembly by 2010, and the CAM was included on that manifest—but the CAM flight was finally deleted for good after STS-114 when NASA pared the 28-flight manifest to 20 flights.

The centrifuge demonstrator dimensions proposed two diameters (9.1 meters and 12 meters), which gave variable Gs (Table 5.1). A kick motor similar to the Hughes 376 spin-stabilizers used on ComSats would be used to start the centrifuge and maintain its rotations. It could also be configured to function as a sleep module for ISS crew.

Table 5.1. Partial-G gravity created by the centrifuge.

RPM	9.1 meters	12 meters
4	0.08	0.11
5	0.13	0.17
6	0.18	0.25
7	0.25	0.33
8	0.33	0.44
9	0.41	0.55
10	0.51	0.69

BEAM

Unlike Bigelow's BA-330 behemoth, the BEAM module (Figure 5.6) destined for the ISS is quite small, and will only add about 2% to the pressurized volume of the orbiting outpost. Four meters in length and 3.2 meters in diameter, the BEAM is roughly cylindrical in shape (a doughnut-shape was considered, but a cylinder was more aligned with NASA's testing requirements). The fabric wall will be just as robust as BEAM's TransHab cousin, consisting of several sets of layers. On the outside, there is a layer comprising sheets of aluminum foil separated from one another by a small space. The space has two functions. One is to make the outer layer act as an extremely effective and lightweight thermal insulation. The second function is that of a Whipple Shield (see sidebar and Figure 5.7), which will vaporize micrometeorites, thereby protecting the inner layers. The next set of layers consists of a thin metal sheet positioned over and separated from a thicker sheet. Their function is similar to the thermal insulation layer, but they also protect the contents of the BEAM module from larger micrometeorites. The wall's inner layers comprise several sheets of Vectran, a robust polymer. These layers protect against external and internal penetration. Despite their impressively rugged design, the walls weigh only about 25 kilograms per square meter, compared to about 110 kilograms per square meter for a similar-sized ISS compartment built using conventional construction. Testing has demonstrated that the BEAM's walls will be at least as resistant to radiation and micrometeorites as the rest of the ISS but with the advantage that, unlike metal, high-energy cosmic rays will pass through without forming a shower of secondary high-energy X-rays.

5.6   BEAM module. Courtesy: Bill Ingalls/NASA

5.7   Example of a Whipple Shield. This one was used on NASA's Stardust probe. Courtesy: NASA

Whipple Shield

Spacecraft shielding must protect against meteorites and orbital debris which have very different velocities. For example, orbital debris impact velocities in LEO average about 11 kilometers per second—a speed which focuses an extraordinary amount of kinetic energy at the impact point. In contrast, meteorites have even higher velocities, averaging about 20 kilometers per second and reaching velocities as high as 70 kilometers per second. Protecting occupants ensconced in orbiting habitats against these hypervelocity projectiles is a tough ask of any aerospace engineer. One way to shield against these impacts is to increase the thickness of the spacecraft wall so the wall remains intact after the impact. The problem with this approach is that this significantly increases the weight of the spacecraft. A smarter method is to use the Whipple Shield concept. Named after Fred Whipple, the Whipple Shield was proposed in the 1940s as a meteorite shield for spacecraft. The shield comprises a thin, aluminum sacrificial wall mounted at a distance from a rear wall. The function of the first layer is to break up the projectile into a cloud of material containing both projectile and layer debris. This cloud expands while moving across the gap, resulting in the debris momentum being distributed over the rear layer. Since the Whipple Shield results in a significant weight reduction over a single plate, it's not surprising the idea is still in use today.

Berthing mechanism

While the BEAM technology is developed and the launcher has been decided, there is still the matter of the berthing mechanism. For those who are not aerospace engineers, it is worth highlighting the difference between berthing and docking. Docking is when an incoming spacecraft makes a rendezvous with another spacecraft and flies a controlled (collision) trajectory to align and mesh interface mechanisms. The spacecraft docking mechanisms typically enter what is called soft capture, followed by a load attenuation phase, and then the hard docked position which establishes an air-tight connection between spacecraft. Berthing, in contrast, is when an incoming spacecraft—or module in the case of the BEAM—is grappled by a robotic arm and its interface mechanism is placed in close proximity to the stationary interface mechanism. Not surprisingly, given its myriad elements, the ISS has a number of berthing and docking mechanisms: in fact, the ISS has become a veritable test bed for berthing and docking mechanisms, and the interfaces have taken many different forms, some pressurized to join modules where the crew live and work and some unpressurized to join the primary structural elements together. The interface similar to the one that will be used to berth the BEAM is the Common Berthing Mechanism (Figure 5.8), created by Boeing.

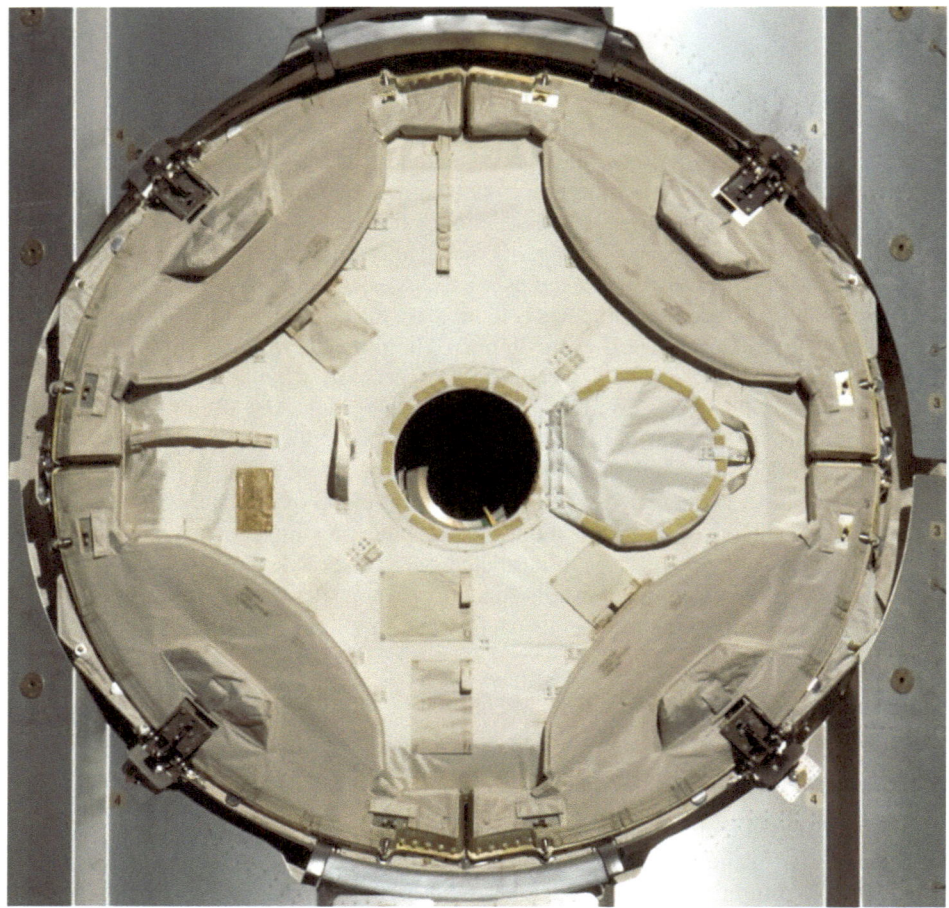

5.8   Common Berthing Mechanism. Courtesy: NASA

This mechanism type, which was used repeatedly for the US segment, is made up of two mating halves: an Active Common Berthing Mechanism (ACBM) and a Passive Common Berthing Mechanism (PCBM) (Figure 5.9).

The generic features of the ACBM include coarse alignment guides, ready-to-latch indicators, capture latches, strike plates, fine alignment pins, and powered bolts, while the generic features of the PCBM include coarse alignment guides, capture fitting, thermal equalization standoffs, fine alignment sockets, and power bolt nuts. Here's how the BEAM berthing will work (the approach sequence of events is explained in Chapter 6). When the PCBM is brought into alignment with the ACBM (through robotic manipulation of the incoming BEAM), the coarse alignment guides will provide coarse alignment by constraining roll and translation. As the incoming PCBM is translated towards the ACBM, the coarse alignment guides will interact with each other and with the ready-to-latch indicators.

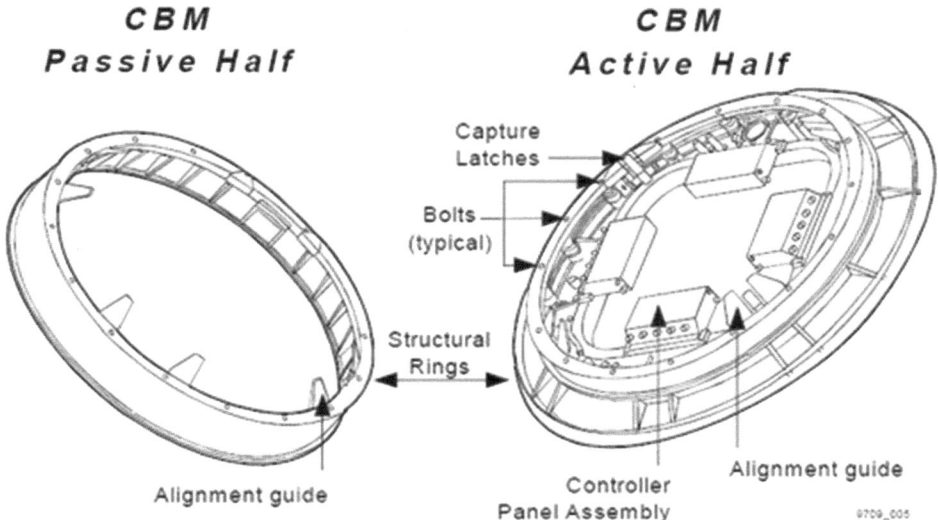

5.9   Passive and common berthing mechanisms. Courtesy: NASA

When the ready-to-latch indicators are triggered by the PCBM alignment guides, the capture latches will be deployed and draw the PCBM into alignment, which will be accomplished when the ACBM alignment pins are seated in the PCBM alignment sockets. Once alignment is achieved, structural attachment will be completed by advancing first all 16 and then in groups of four the 16 powered bolts on the ACBM into the 16 nuts on the PCBM.

The company contracted to build the berthing mechanism is Sierra Nevada Space Systems of Louisville, Colorado. Sierra Nevada solidified its docking and berthing technology by being a major subcontractor on the Orbital Express program, providing a system that captured and docked two spacecraft together on orbit to allow for remote servicing such as refueling and replacement of outdated and expended components. The company then leveraged that mechanical systems experience into becoming the go-to supplier for the industry standard PCBM, required for spacecraft such as the Orbital Cygnus Advanced Manoeuvring Vehicle and, more recently, the BEAM (Sierra Nevada received nearly US$2 million from NASA to build the PCBM). Once the mechanism is finished, Sierra Nevada will bring it to Bigelow's North Las Vegas factory and install the hardware on the BEAM.

Habitat evaluators

The prospect of a BEAM being attached to the ISS created a new job opportunity: an inflatable habitat evaluator (see sidebar). Bigelow Aerospace advertised the job at the end of 2013, asking for applicants to come to its Las Vegas facility and pretend to be astronauts for a few hours. The successful candidates will spend periods of time in a closed-volume spacecraft simulation chamber where they will eat, sleep, and exercise while being monitored. In addition to being given daily tasks and schedules, the habitat evaluators will produce detailed daily reports on their activities and on their interactions with other crew-members. Bigelow Aerospace hopes the reports will help the company quantify, evaluate, and optimize crew systems for use in later iterations of its modules, such as the BA-330, of which more later. It also hopes to glean important information about process efficiencies, program quality, and the psychological, social, and environmental factors in space-craft crews. Applications were restricted to US citizens and to those holding a BS or MS in social, psychological, behavioral, biological, nursing, engineering, or human factors sciences. Bigelow will use its standard BA-330 module (capable of holding a crew of six) in the studies.

Closed-Volume Spacecraft Simulation Crewmembers

This is a part-time position. All qualified applicants will receive consideration for employment without regard to race, color, religion, sex, or national origin.

Duty location: North Las Vegas, Nevada

Bigelow Aerospace seeks mature, well-adjusted adult individuals with backgrounds in the social, psychological, behavioral, biological, nursing, engineering, or crew systems sciences for astronaut-in-space simulation studies.

Qualifications: Demonstrated expertise in detailed report writing with requested education background below.

US citizens and permanent residents only.

Responsibilities: The successful candidates will be expected to spend 8, 16, or 24-hour periods in a closed-volume spacecraft simulation chamber. Candidates will live (eat, sleep, and exercise) inside the chamber for defined periods of time and will be monitored continuously. Successful candidates will be given structured daily tasks and schedules and will be expected to produce detailed daily reports on their activities and on their interactions with other crewmembers. The candidates will implement Bigelow Aerospace programs for quantifying, evaluating, and optimizing crew systems, including process efficiencies, program quality, and reporting on psychological, existential, social, and environmental factors in spacecraft crews.

Education: BS or MS in Social, Psychological, Behavioral, Biological, Nursing, Engineering, or Human Factors Sciences.

6

The Space Taxi Race

First there was *Genesis I*, then *Genesis II*, and, if all goes well, there will be a Bigelow Expandable Activity Module (BEAM) attached to the International Space Station (ISS) by 2015. Each mission marks an incremental step in the evolution of Bigelow Aerospace's plan to launch a commercial station comprising linked Bigelow BA-330 modules, perhaps as early as 2016. Will it happen? Probably. After all, there are plenty of launch vehicles capable of lofting BA-330s into low Earth orbit (LEO). But launching clients? Well, that's another matter. Bigelow is not in the business of building launch vehicles, which is why the company must rely on companies which are part of NASA's Commercial Crew Programs (CCPs), such as Boeing and SpaceX. The CCP (see Appendix I) was established to encourage commercial operators to develop man-rated vehicles capable of ferrying astronauts to LEO. Without at least one of these companies succeeding, any chance of a Bigelow Space Station being flown any time soon are slim to none because there won't be any way to ferry clients to the station. So, before looking ahead to the possibility of Bigelow's space station becoming operational, it's worth taking time to understand the CCP and the vehicles involved.

COMMERCIAL CREW CONCEPT

To begin with, it's worth highlighting that the CCPs were predicated on the concept that NASA and the ISS could function as a sort of anchor tenant in LEO. But with new ventures such as a Bigelow station in the pipeline, there could soon be a second destination for the new crop of commercial vehicles. This is good news for some companies looking to make money by flying its vehicles. For example, Boeing has stated that the addition of a second destination such as Bigelow's station could support its business case for the company's CST-100 capsule (Figure 6.1). And Boeing isn't the only company developing transportation systems. SpaceX, with its Falcon family of rockets (Figure 6.2), has been flying ISS missions for some time now.

Boeing and SpaceX have signed agreements with Bigelow for transportation services to a Bigelow station, and the launch of Bigelow's BEAM module on SpaceX's Falcon 9 booster in 2015 may well be a harbinger of greater and grander things to come. For example, as

© Springer International Publishing Switzerland 2015   99
E. Seedhouse, *Bigelow Aerospace: Colonizing Space One Module at a Time*,
Springer Praxis Books, DOI 10.1007/978-3-319-05197-0_6

6.1   A NASA astronaut prepares to enter Boeing's CST-100 capsule. A ride on this will cost US$36.75 million. That's more than US$10 million more than a seat on SpaceX's Dragon. Courtesy: NASA

many in the commercial space industry have noted, the addition of an inflatable module similar to the BEAM to a Dragon 2.0 significantly increases the space and capability of the capsule to serve as a Mars transfer vehicle and/or surface habitat. And if you add the Falcon Heavy to the interplanetary exploration equation, you have all the pieces needed for an affordable vision of inner system exploration—a subject discussed in Chapter 8.

Bigelow has said his company plans to have two BA-330 modules ready to launch by 2016, although he hasn't specified when the modules, which are still in development, might actually fly. That's because, while the modules and subsystems may be ready by 2016, getting them into LEO may prove to be more challenging. In the past, he has said either SpaceX or United Launch Alliance (ULA) might provide the ride, emphasizing his company won't be launching any BA-330s until there is a commercially available, astronaut-carrying spacecraft to take his customers there. It was this lack of a commercial manned space capability that forced Bigelow Aerospace to lay off nearly two-thirds of its employees in 2011. Fast forward to 2014, and little has changed, although the company now has about 120 employees (after the 2011 cull, it only had 51). So what is the delay getting these commercial vehicles ready? In short, funding and the politics tied to the CCP.

6.2   SpaceX's Falcon 9 launch vehicle with the Dragon capsule perched on top. Courtesy: SpaceX

COMMERCIAL CREW PROGRAMS

Since the end of the Shuttle program in July 2011, the US has lacked a domestic capability to transport crew and cargo (until SpaceX came along) to the ISS. This has meant NASA has been relying on the Russian Federal Space Agency (Roscosmos) for crew transportation. NASA pays exorbitant fees for the privilege. For example, between 2012 and 2017, NASA will pay Roscosmos US$1.7 billion to ferry 30 NASA astronauts and international partners to and from the ISS at prices ranging from US$47 million to more than US$70 million per astronaut. SpaceX, in contrast, is offering seats on board its Dragon for just US$26 million. After 2017, NASA hopes to ferry its astronauts to the ISS via one or more American spaceflight companies. To achieve that goal, the agency is working with three companies—Boeing, SpaceX, and Sierra Nevada Corporation (usually referred to as Sierra Nevada)—to develop commercial crew transportation capabilities using a combination of funded Space Act Agreements (SAAs; these are beneficial because they allow sharing of development costs among the partners) and procurement contracts (see Appendix I). Keeping a close eye on how spacecraft development is progressing is Bigelow Aerospace, which is hoping Boeing and/or SpaceX can man-rate their vehicles sooner rather than later. Bigelow is also praying that Congress keeps the money flowing so NASA can fund these companies because, while Boeing, SpaceX, and Sierra Nevada are responsible for developing the vehicles, they rely heavily on NASA funding. At the same time, NASA must ensure that its partners' launch systems and spacecraft meet the agency's safety and operational requirements. In 2012, the three companies achieved a state of maturity approximate to a Preliminary Design Review (PDR) prior to NASA's award of the next round of SAAs and set a schedule for achieving a company-defined Critical Design Review (CDR) of their systems by mid-2014. But, while NASA's commercial partners are making steady progress developing their spacecraft, NASA faces political obstacles that may prevent it from fully funding these partners. If that happens, transporting astronauts in commercially supplied vehicles may be delayed beyond 2017. Such a delay would have a negative impact on Bigelow, leaving the company with no access to its orbiting BA-330s, and leaving its sovereign customers stranded on Earth.

COMMERCIAL CREW PROGRAMS STALLED

NASA's problems include unstable funding, alignment of cost estimates with the program schedule, challenges in providing timely requirement and certification guidance to commercial partners, and ongoing coordination with the Federal Aviation Administration (FAA). Failure to address these challenges could delay the availability of commercial crew transportation services and extend US reliance on Russia for transporting US crew to the ISS and Bigelow's BA-330s. That's not a position Bigelow Aerospace wants to find itself in and it's not a position that NASA—or its international ISS partners—wants to find itself in either. In the eyes of some in the commercial spaceflight industry, the agency has been coasting for a long time, kept alive by the distant memory of the lunar landings and less spectacular missions such as the unmanned probes to Mars. Sometimes it seems NASA has become a job-protection racket, spending millions of dollars on programs and ideas

6.3   The aging, and very expensive, Soyuz. Courtesy: Wikimedia

that always seem to get cancelled, although more often than not it is Congress and not the space agency that cancels the programs. The result? In 2011, after having spent tens of billions on the ISS, NASA no longer had a way to get there, unless it paid for through the nose for tickets on board the Soyuz (Figure 6.3).

The long-term solution to the problem of ferrying crews to the ISS has been known to NASA and the private space industry for a long time: figure out how the agency can get out of the way and help private companies take the next step by commercializing space. In a nutshell, NASA decided that the best way to get Americans and American companies back into space was for the government to partner with private enterprise, which is what it did via a collection of CCPs. The goal was to provide technical expertise and legal author-ity for ambitious entrepreneurs to spend their own money on endeavors that would not only re-establish American supremacy in space, but also get started on exciting long-range projects, including private space stations, and perhaps even permanent bases on the Moon, and on Mars. So what went wrong?

Well, the CCP isn't derailed. Just stalled. One problem is that, between 2011 and 2013, the CCP received only 38% of requested funding, bringing the aggregate budget gap to US$1.1 billion when comparing funding requested to funding received. In addition, although NASA's partners have completed their preliminary spacecraft designs, by early 2014, NASA managers had yet to develop a life cycle cost estimate showing the antici-pated costs of the program throughout its life from preliminary design to the end of opera-tions. Without this cost estimate, it is difficult for NASA to calculate how much funding is required each year because costs fluctuate over time. And, if NASA can't provide accurate information, costs increase and schedule overruns are the likely result, all of which will delay the BA-330. NASA wants at least one of the spacecraft it is funding under the CCP to be operational by the end of 2017, but with all this funding uncertainty, that date could slip. What may be really frustrating is that, as the agency struggles to execute ambitious programs on increasingly tight budgets, the main beneficiary will most likely be the crisis-prone—and expensive—Russian space agency Roscosmos, which will reap yet another financial windfall as a result of America's confused space policy.

NASA's Commercial Crew and Cargo Program

NASA's Commercial Crew and Cargo Program was established to invest financial and technical resources to catalyze efforts within the commercial sector to develop and demonstrate safe, reliable, and cost-effective space transportation capabilities. The program manages Commercial Orbital Transportation Services (COTS) partnership agreements with US industry totaling US$800 million for commercial cargo transportation demonstrations. It also oversees the Commercial Crew Development (CCDev), a NASA investment funded by US$50 million of the American Recovery and Reinvestment Act (ARRA) funds, to stimulate efforts within the private sector that aid in the development and demonstration of safe, reliable, and cost-effective space transportation capabilities. CCDev-2 commercial partners include Blue Origin, Boeing, SNC, and SpaceX. In essence, the Commercial Crew and Cargo Program is a multiphase strategy that doles out funds to companies to develop solutions for crew transportation to LEO.

Worse, ISS is solely reliant on a single transportation system—Soyuz—to get astronauts up and down. There is no backup. But, rather than giving American companies more money to fill the spaceflight gap, Congress decided to spend US$3 billion per year on the Space Launch System (SLS) and Orion (Figures 6.4 and 6.5) capsule, which won't fly with astronauts until 2021. At the earliest. In comparison, commercial crew spending is stuck at just over US$500 million annually (by early 2014, NASA had spent nearly US$2 billion to Russia for crew services!). Nobody has ever said politicians are good at mathematics, but such a policy makes no sense at all.

So where does this leave Bigelow? Well, additional schedule delays could push the start of commercial flights towards 2020, which is four years before the ISS is scheduled for decommissioning. And, since the ISS is being decommissioned in 2024, planning for what comes next needs to get started now. The first step in figuring out what should be built after ISS is figuring out who the potential customers are and

6.4 NASA's Space Launch System. Courtesy: NASA

6.5 Orion. Courtesy: NASA

Table 6.1. ISS operating costs in 2013.

NASA s budget estimates for space operations	
ISS operations and management	US$1.4935 billion
ISS research	US$229.3 million
Crew and cargo transportation	US$1.2848 billion
Total	US$3.0076 billion

identifying needs, which is something Bigelow has already done to some degree. If the next station or stations are commercial, the potential for opening up new commercial markets for space activities increases, especially if costs are significantly lower than the ISS (Table 6.1).

While NASA struggles with its budget and over-the-horizon planning, China is busy preparing its own multi-module space station, the first element of which will be launched in 2018. And, taking a page out of Bigelow's business plan, China is eagerly recruiting international partners to fly astronauts and experiments on board the facility. If the Chinese orbit their station before Bigelow can orbit his, they will have taken an important step towards commercializing LEO, perhaps to the detriment of Bigelow. If this happens, Bigelow may have to hope he can sell one of his stations to NASA after the ISS is decommissioned. Chances are that a station purchased commercially would be affordable and less costly to run and supply, although it remains to be seen whether such a station can replicate the research capabilities that the ISS possesses.

COMMERCIAL CREW OPTIONS

"NASA announces US$1.1 billion in support for a trio of spaceships." That was one of the headlines on August 3rd, 2012, when NASA announced it had committed US$1.1 billion over the next 21 months to support spaceship development efforts by a line-up of companies with the aim of having American astronauts flying on American spacecraft within five years. The companies included aerospace juggernaut, Boeing, which received US$460 million in the Commercial Crew Integrated Capability (CCiCap) outlay; SpaceX, which received US$440 million; and Sierra Nevada, which received US$212.5 million. Also in the space taxi hunt were Orbital Sciences Corporation (OSC), Blue Origin, and ATK Aerospace Systems.

NASA's CCiCap (see Appendix I) announcement heralded the next phase of the agency's commercial spaceflight effort. In short, CCiCap called for Boeing, Sierra Nevada, and SpaceX to take their design and testing program through a series of milestones by May 2014, with optional milestones possibly leading to crewed demonstration flights in later years. The goal of the program was to have at least one commercial space taxi ferrying astronauts[1] to and from the ISS by 2017. The companies said they were confident they could meet or beat the schedule—provided they continued to receive NASA support. It was good news for NASA Administrator, Charles Bolden, whose agency was—and still is—at the mercy of a sole rocket provider charging up to US$70 million a seat.

CCiCap is the third phase of NASA's CCP. In earlier phases, Boeing, SpaceX, and Sierra Nevada had received hundreds of millions of dollars in NASA support. While SpaceX is rapidly upgrading its Dragon capsule to manned capability, Boeing is working on its CST-100, and Sierra Nevada is testing its Dream Chaser spaceplane (Figure 6.6), which looks like a miniature version of the Space Shuttle. In a statement, Elon Musk, SpaceX's CEO and chief designer, hailed the CCiCap award as "a decisive milestone in human spaceflight" that would set "an exciting course for the next phase of American space exploration". Boeing's statement struck a similar tone, with John Elbon, Boeing vice president and general manager of space exploration, announcing "Today's award demonstrates NASA's confidence in Boeing's approach to provide commercial crew transportation services for the ISS. It is essential for the ISS and the nation that we have adequate funding to move at a rapid pace toward operations so the US does not continue its dependence on a single system for human access to the ISS". Bigelow probably couldn't have agreed more.

The CCiCap funding was partly thanks to the success of SpaceX because, before Dragon's first flight, the plan for supporting numerous competitors had been in danger. Some in Congress, such as NASA Appropriations Chairman Frank Wolf of Virginia, had been pressuring the agency to select a single provider to save money. But, following Dragon's successful ISS flight, Wolf was persuaded to let the competition continue.

---

[1] Incidentally, NASA astronauts will fly as spaceflight participants on board commercial spacecraft. That's the ruling of the FAA Office of Commercial Space Transportation, which decided NASA astronauts didn't meet the definition of "crew" because that definition requires them to be employees of the licensee or subcontractor licensee. NASA astronauts are neither, so they will be flying as spaceflight participants—at least under the current regulations.

6.6   The Dream Chaser flight vehicle is prepared for 100 kilometer per hour tow tests on taxi and runways at NASA's Dryden Flight Research Center at Edwards Air Force Base in California. Courtesy: Ken Ulbrich/NASA/Sierra Nevada/Wikimedia

This, for NASA, was a good thing, because the agency didn't want to be dependent on a single means of getting its astronauts into orbit. They had tried this strategy with the Space Shuttle and it hadn't worked out (the Shuttle program was down for more than five years during its lifetime for investigations after accidents).

Whichever company or companies NASA selects for carrying crew, the new vehicle will go beyond a Shuttle replacement in one important way. Since the ISS was first occupied more than a decade ago, there has always been a Soyuz capsule berthed to provide a rescue capability. The reliance on the Russians for the lifeboat service was because the Space Shuttle never had the ability to stay at the station for more than a week or so. However, all the new vehicles are designed for an orbital life of several months and, with room for up to seven crewmembers, they are larger than the claustrophobic confines of the three-passenger Soyuz. Using one or more of these commercial capsules as the ISS's lifeboat service will be a game-changer not only because it will eliminate NASA's dependence on the Russians (with whom relations can be rocky), but also because these new spacecraft will allow an expansion of ISS crew capacity, currently limited by lifeboat size.

The August 2012 CCiCap announcement, which was the final-phase development funding under NASA's CCP, established Boeing and SpaceX as the clear frontrunners in the space taxi race, with Sierra Nevada waiting in the wings as the fallback option in case one of the other two falters. If Boeing and SpaceX meet their NASA-approved milestones in the 21-month CCiCap performance period, their designs will undergo a CDR.

If the CDRs go well, construction can begin. The three CCiCap winners said they could stage their first unmanned demonstration flights by 2015 or 2016, although construction and flight tests are not funded under the initial CCiCap awards. Boeing officials said the CST-100 could be ready to conduct its first manned flight by 2016.

What NASA was looking for was a diverse range of technical approaches, a strong business approach, and spacecraft development versus launch vehicle development. This last requirement made sense, since the US has plenty of launch vehicle development expertise and experience but very little experience developing crew-carrying spacecraft. When it came to selecting the winning three, the Boeing and SpaceX proposals stood out from the rest thanks to their high ratings in technical and business factors, which is good news for Bigelow.

BOEING

Profile

Spacecraft: CST-100
Type: Capsule with service module
Crew capacity: Seven
Launch vehicle: Atlas V (ULA)
CCiCAP funding: US$460 million
CCiCAP term: 21 months
Previous CCDev funding (including optional milestones): US$130.9 million (Boeing), US$6.7 million (ULA)
Total CCDev and CCiCAP funding (if all milestones met): US$590.6 million (Boeing), US$6.7 million (ULA)

CST-100

Rather than using heritage technology from the lifting body program, Boeing is advancing plans for its new capsule-based spaceship. The CST-100 capsule is a spacecraft design proposed by Boeing in collaboration with Bigelow Aerospace as their entry for the CCDev program. As with all the other competitors in the space taxi race, the CST-100's primary mission is transporting crew to the ISS and to private space stations such as Bigelow's. At first glance, the gumdrop-shaped capsule looks similar to the Apollo and Orion, the latter a spacecraft being built for NASA by Lockheed Martin.

When complete, the CST-100 (incidentally, the number "100" stands for 100 kilometers, the height of the Kármán line, which defines the boundary of space) will be larger than the Apollo command module but smaller than the Orion. Capable of ferrying crews of up to seven thanks to a generous habitable interior and the reduced weight of equipment needed to support an exclusively LEO configuration, the CST-100 is designed to remain on orbit for up to seven months and reusability for up to 10 missions. Although the initial launch vehicle for the CST-100 will most likely be the Atlas V, the spacecraft will be compatible with multiple launch vehicles, including the Delta IV and the Falcon 9.

In common with Sierra Nevada, Boeing is also the recipient of earlier NASA funding. In the first phase of the CCDev program, the company was awarded US$18 million for preliminary development of the spacecraft—a sum that was followed by another US$93 million in the second phase for further development. Although an industry juggernaut, Boeing was still reliant on external funding for development of the CST-100, the company stating in July 2010 that the capsule could only be operational in 2015 with sufficient near-term approvals and funding. Boeing also indicated they would proceed with development of the CST-100 only if NASA implemented the commercial crew transport initiative announced by the Obama Administration in its FY11 budget request. In fact, Boeing's business case was reliant not only on continued NASA funding, but also on the existence of a second destination, hence the partnership with Bigelow.

While the CST-100 is sometimes confused with its big-budget cousin, the Orion, the Boeing vehicle has no Orion heritage, its design drawing mostly upon Boeing's experience with Apollo, the Shuttle and ISS programs, and the Department of Defense's Orbital Express project.

Measuring 4.5 meters across at its widest point and standing 3.1 meters high, the CST-100 sports a four-engine "pusher abort system" rather than an escape tower as used in the Mercury and Apollo programs and is designed to land on terra firma, although it can also support a water landing in the event of an abort. For docking with the ISS, the vehicle will use the Androgynous Peripheral Attach System (APAS) while, for re-entry, the Boeing Lightweight Ablator (BLA) heat shield will protect the crew.

Work on CST-100 design, manufacture, testing, and evaluation is well underway and moving at a rapid pace thanks to Boeing pulling in proven technology. For example, the CST-100 utilizes the technology developed to build the Apollo-era heat shield and the Space Shuttle's thermal protection system, as well as autonomous rendezvous and docking gear developed on the Pentagon's experimental satellite-refueling Orbital Express mission. The CST-100 also uses flight computers currently in use on the Boeing-built X-37 spaceplane. Given that Boeing is working on a fixed-price development, using mature designs and drawing upon flight-proven hardware not only makes sense financially, but also gives them an edge on the competition.

Some of the more notable tests performed by Boeing in its development of the CST-100 include drop tests to validate the design of the air bag cushioning system and the capsule's parachute system. The air bags, which are deployed by filling with a mixture of compressed nitrogen and oxygen, are located underneath the CST-100's heat shield, which is designed to be separated from the capsule while under parachute descent at about a 1,500-meter altitude. In September 2011, Boeing conducted drop tests in the Mojave Desert in California, at ground speeds between 16 and 48 kilometers per hour to simulate cross-wind landing conditions (Bigelow Aerospace built the mobile test rig and conducted the tests). The drop tests were followed by parachute tests in April 2012, when Boeing dropped a CST-100 mock-up over the Nevada Desert at the Delamar Dry Lake near Alamo, Nevada, successfully testing the craft's three main landing parachutes (Figure 6.7). Mr. Bigelow was impressed, noting in a press statement that "If astronauts had been in the capsule during these drop tests, they would have enjoyed a safe, smooth ride … further proof that the commercial crew initiative represents the most expeditious, safest, and affordable means of getting America flying in space again".

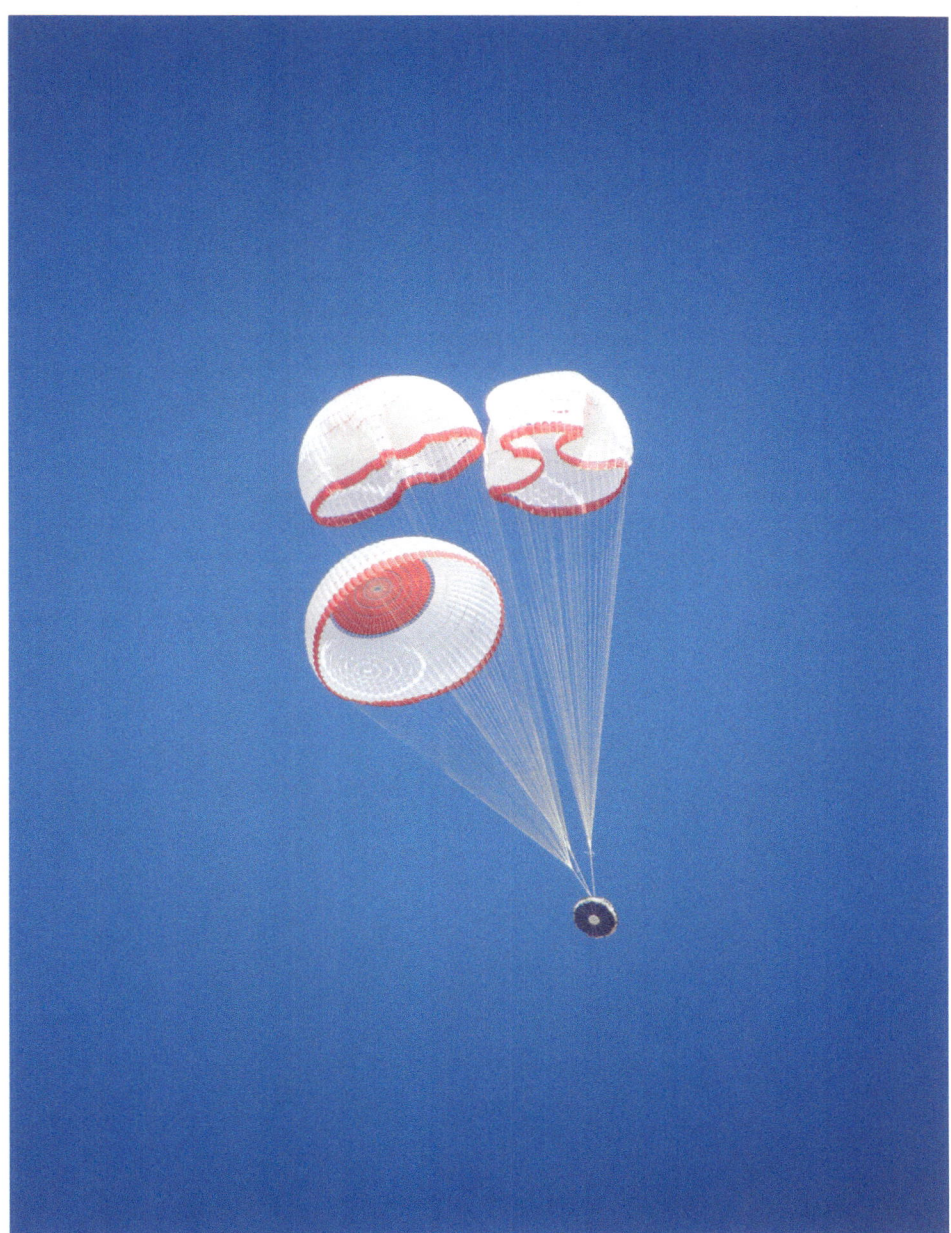

6.7   CST-100 under canopy. Courtesy: Boeing

Meanwhile, manufacturing of the CST-100 continues at the Orbiter Processing Facility-3 at Kennedy Space Center—a facility leased to Boeing through a partnership with Space Florida in October 2011. If all goes well, Boeing's private spaceship may be ready to carry astronauts to the ISS, Bigelow habitats, and other LEO locales by 2017. Casting an eye towards the business case, Boeing expects markets to materialize but, given the unpredictability of the private commercial spaceflight market, the company is hedging its bets by having a low, medium, and high business model. Those business models consider existing partnerships such as the one Boeing has with Bigelow Aerospace and a partnership with space tourism provider Space Adventures, which intends to sell unused seats on the CST-100 for flights to LEO. For Bigelow's clients, the ride to their orbiting inflatable space station will be very comfortable (Figure 6.8).

As you can see in Figure 6.8, the spacious spacecraft has room for two rows of crew seats and cargo storage, including a freezer used to transport science experiments to and from the station. Although it is designed to carry a crew of seven, one configuration seats five, trading two additional seats for more cargo room. The flight controls, which are mounted on a console suspended above the front-row seats, uses Shuttle-era switches and hand controllers, augmented by touch-panel digital displays. A window located forward of the control console offers the pilots a view, with additional windows to either side, and a side hatch allows entry and exit (the overhead hatch leads into the station after docking. One element missing, at least for now, is a toilet, or waste containment system, as NASA prefers to refer to these things.

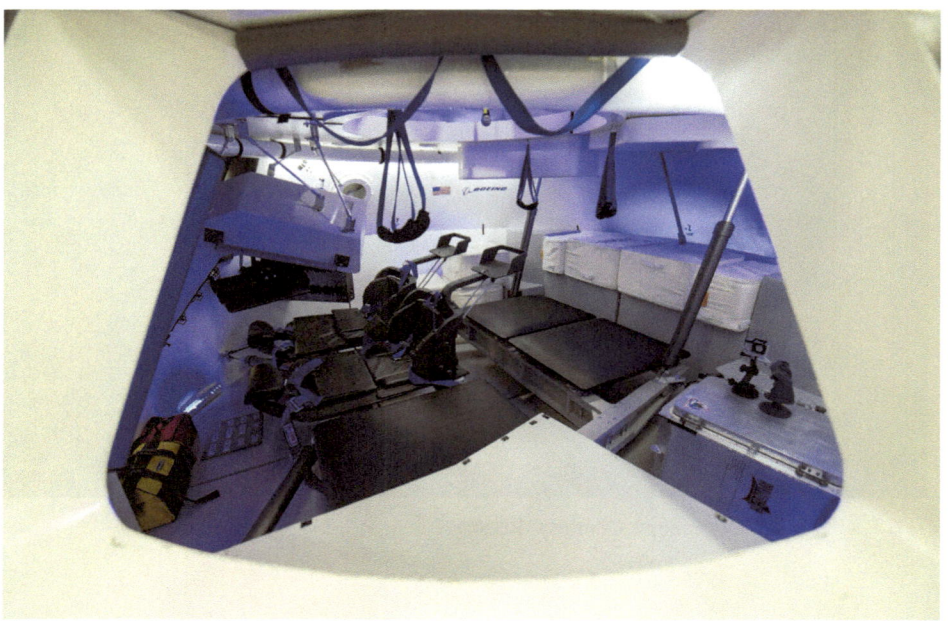

6.8 Interior of CST-100. Courtesy: Boeing

Table 6.2. Boeing CST-100 B-roll sequence of events.

Stage	Mission elapsed time	Event
1	00:15:00	CST-100 second stage to LEO
2	00:23:00	CST-100 separation
3	00:31:03	CST-100 approaches Bigelow Orbital Space Complex
4	00:36:25	CST-100 docks with Bigelow Orbital Space Complex
5	01:17:18	CST-100 undocks
6	01:32:27	CST-100 service module separates from crew module
7	01:38:38	CST-100 re-enters atmosphere, parachutes deploy, and CST-100 lands

At the time of writing, the CST-100 program is on track to deliver its first flight-design hardware with a first flight tentatively scheduled by the end of 2016, although Boeing is looking at ways to launch earlier. The mission events are listed in Table 6.2 but, for those who prefer watching a computer-generated launch to the capsule and its docking with a Bigelow habitat, simply go to www.youtube.com/watch?v=Mn_gXEK5XmQ.

SPACEX

Profile

Spacecraft: Dragon
Type: Capsule with service module
Crew capacity: Seven
Launch vehicle: Falcon 9
CCiCAP funding: US$460 million
CCiCAP term: 21 months
Previous CCDev funding (including optional milestones): US$130.9 million (Boeing), US$6.7 million (ULA)
Total CCDev and CCiCAP funding (if all milestones met): US$590.6 million (Boeing), US$6.7 million (ULA)

Falcon 9

SpaceX's Falcon 9 is a two-stage rocket powered by liquid oxygen (LOX) and rocket-grade kerosene (RP-1). Designed from the ground up by SpaceX for cost-efficient transport of satellites to LEO and geosynchronous transfer orbit (GTO), and for sending the Dragon spacecraft, carrying cargo and/or astronauts to orbiting destinations, this is a true "Made in America" rocket, with all structures, engines, avionics, *and* ground systems designed, manufactured, and tested in the US. Designed to one day carry crew, the Falcon 9, with a Dragon spacecraft perched on top, is 48.1 meters tall. Developed from a blank sheet to first launch in just four and a half years (November 2005 to June 2010), the Falcon 9 is capable of producing one million pounds of thrust in a vacuum. The cost? Less than US$300 million. The vehicle features cutting-edge technology and a simple two-stage design to limit separation events; with nine engines (hence the number "9" in the name) on the first stage, the Falcon 9 can still safely complete its mission in the event of an engine failure.

First and second stages

The tank walls are made from an aluminum lithium alloy that SpaceX manufactures using friction-stir welding—the strongest and most reliable welding technique available. Powering the first stage is a cluster of nine SpaceX Merlin regeneratively cooled engines. Connecting the lower and upper stages is the interstage—a composite structure with an aluminum honeycomb core and carbon fiber face sheets. The separation system is pneumatic—a system proven on its Falcon 1 predecessor.

The Falcon 9 second-stage tank is a shorter version of the first-stage tank and uses many of the same tooling, material, and manufacturing techniques—a measure that results in significant cost savings in vehicle production. Powering the upper stage is a single Merlin engine, capable of restart thanks to dual redundant pyrophoric igniters using triethylaluminum-triethylborane (TEA-TEB). The vehicle is a reliable system, in part because it only has only two stages, which limits problems associated with separation events. This reliability is enhanced by the incredibly advanced avionics systems, which features a hold-before-release system. It's a capability required by commercial airplanes, but not implemented on many launch vehicles. Here's how it works: after the first-stage engine ignites, the Falcon 9 is held down and not released for flight until all propulsion and vehicle systems are confirmed to be operating normally; if any issues are detected, an automatic safe shutdown occurs and propellant is unloaded.

Dragon development

Dragon's development, which began in late 2004 when SpaceX started the design of the capsule using its own funding, formed the centerpiece of the proposal SpaceX submitted under NASA's COTS demonstration program. At 3.6 meters in diameter, Dragon is smaller than Boeing's CST-100 and NASA's five-meter-diameter Crew Exploration Vehicle (CEV) because the SpaceX capsule is intended only for short jaunts to the ISS, and not the longer expeditions to the Moon and beyond. The gumdrop-shaped Dragon comprises a blunt-cone ballistic capsule, not that dissimilar to the design of the Soyuz or Apollo capsules, a nose-cone cap that jettisons after launch, and a trunk equipped with two solar arrays. To protect its cargo and crew during re-entry, the capsule utilizes a proprietary variant of NASA's phenolic impregnated carbon ablator (PICA) material. The spacecraft also features standard options such as a docking hatch, maneuvering thrusters—18 of them—and a trunk, which, unlike the rest of the reusable spacecraft, separates from the capsule before re-entry. During its initial cargo and crew flights, the Dragon capsule lands in the Pacific Ocean and is returned to the shore by ship, but eventually SpaceX will install deployable landing gear and use upgraded Super Draco thrusters to permit solid Earth propulsive landings.

In 2005, the year after SpaceX began developing Dragon, NASA began its COTS development program, soliciting proposals for a commercial re-supply spacecraft to replace the soon-to-be-retired Shuttle. SpaceX submitted Dragon as part of its proposal in March 2006 and, six months later, NASA announced SpaceX had been chosen to develop cargo launch services for the ISS. Dragon's development progressed quickly. In February 2009, SpaceX announced that Dragon's heat shield material had passed heat stress tests in preparation for the capsule's maiden launch. Then, in July 2009, the DragonEye, the capsule's primary proximity-operations sensor, was tested during the STS-127 mission, when it was mounted near the docking port of the Space Shuttle *Endeavour* and used while the Shuttle approached the ISS.

Flights

Dragon's first flight took place in June 2010 when a stripped-down capsule, dubbed the Dragon Spacecraft Qualification Unit (initially used as a ground test bed), was launched on top of a Falcon 9. Dragon's first mission was simply to relay aerodynamic data captured during the ascent; the capsule wasn't designed to survive re-entry because, at the time, SpaceX didn't have a re-entry license—a document issued by the FAA. Following the successful flight of the first Dragon, SpaceX had to wait for the FAA to issue a re-entry license, which the agency did in November 2010—the first such license ever awarded to a commercial vehicle. On December 8th, 2010, the Dragon spacecraft launched on COTS Demo Flight 1, which also marked the second flight of the Falcon 9. Unlike the qualification unit, this Dragon was designed to survive re-entry and was recovered.

Capable of lifting payloads of 10,450 kilograms to LEO, and 4,450 kilograms to GTO, the Falcon 9 launched—after several delays—on its maiden flight from Cape Canaveral Air Force Station on June 4th, 2010, at 2:45 pm EDT. Lift-off came 3 hours and 45 minutes into a four-hour launch window because of tests conducted on the rocket's self-destruct system, and a last-second abort caused by a higher-than-expected pressure reading in one of the engines. SpaceX engineers resolved the issue and recycled the countdown to the T-minus 15-minute mark after concluding the engine was in good shape.

If you consider that two-thirds of rockets introduced in the past 20 years have had an unsuccessful first flight, the maiden Falcon 9 launch couldn't be considered as anything other than 100% successful, but Musk and his team of engineers weren't the only ones breathing a sigh of relief as the Falcon 9 sailed above their heads; Bigelow Aerospace, which had been talking with SpaceX with a view to using the Falcon 9/Dragon combination as a means to ferry its clients to LEO, also started feeling a lot more comfortable (SpaceX and Bigelow Aerospace announced their marketing alliance in May 2012). The successful Falcon 9 launch not only boded well for Bigelow's transportation headache, but also vindicated Musk's approach by demonstrating that a small, new company could make a difference. Talking to the media following the launch, Musk acknowledged NASA's help, but also pointed out that the first Falcon 9 heralded the dawn of new era of spaceflight which will be increasingly marked by combined commercial and government endeavors, with commercial companies playing an increasingly significant role.

Given the success of the second Falcon 9 flight, it wasn't surprising when SpaceX proposed combining two COTS flight demonstrations of the Falcon 9/Dragon. It was planned that the combined flight would precede routine re-supply runs to the station under a separate US$1.6 billion fixed-price contract with NASA. To facilitate the previously unplanned ground tests that would be needed to support the combined demonstration flight, the agency boosted its investment in SpaceX by US$128 million in 2011. The combined flight made sense because, at the time, SpaceX was more than two years behind in completing its COTS demonstration flights. In the original plan, SpaceX's second COTS demo had been a five-day mission during which Dragon would approach to within 10 kilometers of the ISS and use its radio cross-link to allow the station's crew to receive telemetry from the capsule and send commands. In the third and final COTS demo, Dragon would berth with the ISS for the first time. If successful in the first-of-a kind mission, SpaceX would collect the remaining payments on the US$396 million contract it had with NASA and then enter into a US$1.6 billion agreement for 11 more flights to the ISS.

For the COTS 2+, SpaceX used the Crew Resupply Services (CRS) Dragon variant, which didn't have an independent means of maintaining a breathable atmosphere for astronauts and instead circulated fresh air from the ISS. The first launch attempt, on May 19th, 2012, resulted in a countdown abort at T−00:00:00.5. Following the abort, SpaceX announced the next attempt would be scheduled three days later at 03:44 EDT or May 23rd, 2012, at 03:22 EDT. The second attempt was successful. After completing a short vertical ascent, Falcon 9 made its roll-and-pitch maneuver to align itself with the correct flight trajectory. Nine minutes and 14 seconds after launch, the vehicle shut down its engine and, 35 seconds later, Dragon was released. Shortly thereafter, Dragon deployed its solar arrays and SpaceX Dragon Control, based in Hawthorne, California, began its list of vehicle status checks to confirm the capsule was in good condition and ready for operations. Dragon's first day in space was dedicated to Far Field Phasing Maneuvers and various test objectives (Table 6.2). As the mission progressed, Dragon was required by NASA to complete a series of tests and meet a number of objectives (Figure 6.9) before being permitted to rendezvous with the ISS: similar maneuvers will most likely be required by vehicles destined for Bigelow's stations, so it's worth discussing them briefly. One of the first tests was to demonstrate Dragon's GPS navigation capability—a task that was completed about an hour into the flight. Shortly after demonstrating its way-finding capability, Dragon deployed its guidance, navigation, and control (GNC) bay door to put the necessary rendezvous instruments in place. DragonEye, the vehicle's rendezvous and navigation instrument suite, included a Light Detection and Ranging (LIDAR) imager for this purpose. Just after being deployed, each of these instruments was activated and underwent its requisite checkouts. On Flight Day 2, Dragon made several engine burns to adjust its orbit in preparation for its rendezvous with the ISS the following day. The burns, referred to as Far Field Phasing or Height Adjust Burns in NASA parlance, were designed to increase the Dragon's orbital altitude. This required the vehicle to target a point 10 kilometers beneath and behind the ISS.

Once Dragon had arrived at a point 10 kilometers below and behind the ISS, teams at each Mission Control Center conducted a poll before permitting Dragon to conduct its Fly-under Maneuver to set it up in preparation for performing the R-Bar maneuver (RBM). Once approved, Dragon performed two burns before entering the ISS's 28-kilometer communications zone in which it could communicate directly with ISS Systems. For relative GPS communication with the ISS, Dragon used a proximity communications link supplemented by the Commercial Orbital Transportation Services (COTS) Ultra High Frequency

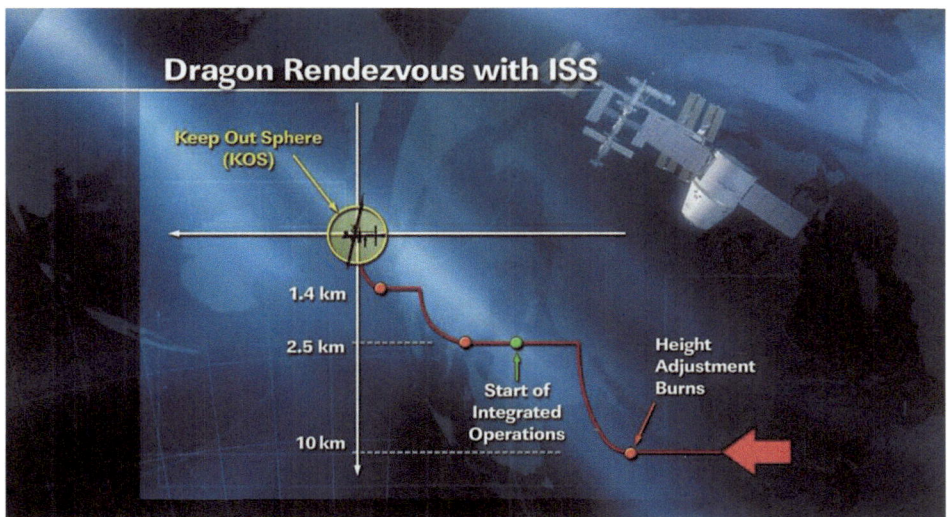

6.9 The Dragon maneuvers depicted here are those made by SpaceX's Dragon capsule when berthing with the ISS. It is anticipated that space taxis en route to Bigelow space stations will follow similar maneuvers. As Dragon approaches the ISS, it makes several engine burns to adjust its orbit to prepare for rendezvous with the ISS. As Dragon gets closer to the ISS, Rendezvous Operations begin, with the spacecraft entering a 28-kilometer communications zone around the ISS. At this time, an Ultra High Frequency (UHF) link between the ISS and Dragon is established, and Dragon uses a proximity communications link for relative GPS communication with the ISS and the UHF communication unit on board the station is used to communicate with Dragon. Using absolute GPS, Dragon makes adjustment burns to reach a position 2.5 kilometers below and behind the station. At 2.5 kilometers, the control teams are polled for the next burn before Dragon is allowed to fire its engines again. At that point, Dragon switches to the relative GPS navigation system for Proximity Operations. As Dragon approaches, it can perform several mid-course correction burns to adjust its trajectory before approach initiation occurs at 1.4 kilometers from the ISS. On board the station, the crew begins monitoring Dragon, ready to take action if necessary. At the 250-meter hold, Dragon completes final adjustments and Mission Controllers are polled before the approach resumes. Once Mission Control has verified Dragon is receiving correct navigation data, the approach is restarted. At 200 meters, Dragon enters the Keep Out Sphere, where the rendezvous is paused as Dragon reduces its relative velocity to zero to give teams time to assess any problems. If everything is okay, Dragon moves to 30 meters and is held again to give Mission Controllers time for another assessment before initiating final approach. Courtesy: NASA

(UHF) Communication Unit (CUCU) aboard the ISS. As it made its close approach, Dragon demonstrated the Relative GPS System (RGPS), which determined the spacecraft's position relative to the ISS.

Once Dragon had passed 2.5 kilometers below the ISS and all operations were complete, the vehicle made a trajectory adjustment to retreat to a distance of 10 kilometers to begin the ISS fly-around. For the fly-around maneuver, the vehicle made several engine burns to cross the ISS's Velocity Vector (V-Bar) flying to a location seven kilometers above the V-Bar before making another maneuver to reduce its velocity. Dragon then passed over the ISS, making one more height adjust burn to increase the distance between itself and the ISS to 10 kilometers. Dragon once again fired its engines to cross the V-Bar one more time, targeting a point behind and below the ISS. The sequence of events took a full day and finally set the stage for rendezvous. The data acquired during the fly-under was reviewed by SpaceX and NASA teams, after which the ISS Mission Management Team made a Go/No Go decision for rendezvous.

At the 250-meter hold, several Dragon systems were checked, including rendezvous navigation systems and LIDAR to demonstrate DragonEye's capabilities. Once Mission Control verified that Dragon's position and velocity were accurate, the approach restarted and Dragon made short engine pulses to re-initiate the rendezvous. Once the spacecraft reached 220 meters on the R-Bar, the ISS crew sent a retreat command to demonstrate one of Dragon's rendezvous-abort capabilities. Once this had been done, Dragon fired its engines to return to the 250-meter hold point—a maneuver it would be required to perform at any stage during the approach if a retreat command was sent. While Dragon completed the retreat operation, Mission Control assessed whether the spacecraft maintained its range from the ISS and that its acceleration and braking performance remained stable.

Once all the maneuvers had been verified, Dragon recommenced its approach and demonstrated the second abort scenario, which required the ISS crew to issue a hold command that initiated a period of station-keeping at 220 meters. Once this had been performed, Mission Controllers verified Dragon's braking performance was nominal and confirmed that the vehicle had stayed within the required range. When all these objectives had been met, Mission Control gave a Go for close approach, at which point Dragon fired its engines and closed in on the ISS, entering the Keep Out Zone while the ISS crew watched closely to make sure there were no problems during the approach. As the vehicle approached the 30-meter mark, Dragon made another hold to give the two Mission Control Centers the opportunity to check the vehicle's status and conduct another Go/No Go poll before allowing Dragon to proceed. After being approved to continue its approach, Dragon crept towards the ISS before coming to a stop just 10 meters from the station. It had reached the Capture Point. When it was verified that Dragon was in the proper position, free drift was initiated and Dragon's thrusters were disabled. The rest of Dragon's rendezvous was performed by the Canadarm under the control of Don Pettit with assistance from Andre Kuipers. The Canadarm captured Dragon and began a carefully choreographed maneuver to place the vehicle above its intended berthing position at the Earth-facing (nadir) Common Berthing Mechanism (CBM) on the Harmony Module. In a mission to a Bigelow inflatable, Dragon will dock without the use of a robotic arm.

Next, four Ready to Latch Indicators were used to confirm that the spacecraft was in the correct position and ready for berthing. Procedures then began to perform first- and second-stage capture of the spacecraft, at the end of which Dragon was secured in place, forming a hard-mate between itself and the ISS (Figure 6.10). This marked the official start of docked operations.

Flight Day 5 began with a new addition berthed to the ISS. One of the first tasks of the day was to pressurize the vestibule between the Dragon and Harmony hatches and check for pressure leaks to make sure the seal between the ISS and the vehicle was tight.

6.10 Dragon berthed at the International Space Station. Courtesy: NASA

Once the leak checks were complete, Mission Control gave the "Go" to open the hatch, at which point crewmembers began the vestibule outfitting task which required the installation of ducts and the removal of equipment needed to bolt Dragon in position. Once both hatches were open, the crew conducted air sampling inside Dragon as part of standard ingress operations. Then, Mission Control gave the green flag for Dragon ingress, allowing the crew to commence cargo transfer operations.

With cargo operations complete, Dragon was closed out, its hatch closed, ducts removed, and control panel assemblies re-installed. Once Harmony's hatch closed again, the leak check procedure was repeated and the vestibule between the two spacecraft was depressurized to prepare for Dragon's unberthing. Once again, the Canadarm was used to grapple Dragon before releasing the vehicle. With Dragon flying free, the Canadarm maneuvered the vehicle to its release position 10 meters from the ISS. At this point, Dragon was in Free Drift Mode (FDM) with all thruster systems inhibited. Dragon's navigation instruments underwent the requisite checkouts to ensure the vehicle was receiving correct navigation data and, with all checks complete, both Mission Control Centers gave the "Go" for release. Then, Dragon vehicle was ungrappled, the Canadarm retreated, Dragon reactivated its thrusters, and recovered from FDM. After performing three engine burns to leave the vicinity of the ISS, Mission Control Houston verified that Dragon was on a safe path away from the station.

As it moved away from the ISS, Dragon closed its GNC control bay door to protect the instruments during re-entry. Four hours after release, Dragon was at a safe distance from the station and fired its engines before making the de-orbit burn taking it on a trajectory to re-enter Earth's atmosphere. Twenty minutes after the de-orbit burn, Dragon hit the entry interface and began to feel the effect of Earth's atmosphere. During re-entry, Dragon's PICA-X heat shield withstood temperatures up to 1,600°C. During the entry phase, Dragon used its Draco thrusters to stabilize its position and control its lift to target the landing. About 10 minutes before splashdown, at an altitude of 13.7 kilometers, Dragon opened its dual drogue chutes, which triggered the main chute opening command, which occurred at an altitude of three kilometers. Descending under its main chutes, Dragon slowed to its landing speed of 17 to 20 kilometers per hour and splashed down, landing about 450 kilometers off the California coast. Experts immediately hailed the historic flight as a new era for private spaceflight: Dragon had joined Russia's Progress, the ESA's Automated Transfer Vehicle (ATV), and the Japan Aerospace Exploration Agency's H-II Transfer Vehicle as regular station suppliers. However, Dragon was the only provider in the international line-up with the capability of returning a significant cargo of research samples and equipment in need of refurbishment to Earth, a critical part of future science activities planned for the six-person orbiting science laboratory and for Bigelow's sovereign clients.

WINNING THE SPACE TAXI RACE

Who will win the space taxi race? It's difficult to say but it will probably be one of the companies awarded CCiCap funding. As it stands in mid-2014, SpaceX's Falcon 9–Dragon combination represents the only all-American launch solution on the table, designed and built in California, tested in Texas, and launched in Florida. For a US workforce desperate for jobs, and a country looking for something to celebrate, it doesn't get much better than that. The company also happens to be at the forefront of the space taxi race, investing its own resources to build what everyone agrees is necessary to establish a permanent future in space: a fully reusable space transportation system. If all goes well, that system could be in service by 2017. All might not go well, of course. Delays are possible, even probable, especially in this industry, as SpaceX and Bigelow know only too well.

7

Bigelow's Space Station

7.0 Artist's rendering of a future Bigelow station showing a Boeing CST-100 about to dock. Courtesy: Boeing

"Today we're demonstrating progress on a technology that will advance important long-duration human spaceflight goals. NASA's partnership with Bigelow opens a new chapter in our continuing work to bring the innovation of industry to space, heralding cutting-edge technology that can allow humans to thrive in space safely and affordably."

NASA Deputy Administrator Lori Garver announcing on January 16th, 2013, that a Bigelow Expandable Activity Module will be added to the International Space Station in 2015 to test expandable habitat technology

© Springer International Publishing Switzerland 2015
E. Seedhouse, *Bigelow Aerospace: Colonizing Space One Module at a Time*,
Springer Praxis Books, DOI 10.1007/978-3-319-05197-0_7

7.1   SpaceShipOne. Courtesy: Wikimedia

To appreciate the significance of Lori Garver's announcement, it's necessary to go back to October 2004, when a flimsy rocket dubbed SpaceShipOne (SS1) (Figure 7.1) glided down to the Mojave Desert following a suborbital jaunt into space. Not only did the flight win the X-Prize and a cool US$10 million, it was also a milestone in the commercial development of space because the feat was achieved without accepting any government dollars. It was an accomplishment that Mr. Bigelow (who was among those watching the flight) could relate to because, just a few years before SS1's epic flight, Bigelow had announced his intention to get into the commercial space race. That was in 1998 when the concept of commercial spaceflight was alien to just about everybody, so very few paid attention, and even fewer gave him a chance. But, less than two years after the X-Prize winning flight, Bigelow's first creation was orbiting Earth, thereby ensuring that the development of space will never be the same.

TRANSFORMING SPACE

Behind the secure confines of a 50-acre space campus in North Las Vegas, Bigelow quietly got to work investing millions of dollars of his fortune into his aerospace company. Over the years, in a facility surrounded by fences, gates, cameras, and an imposing security force made up of ex-military types, the property tycoon stealthily charged towards the future developing innovative inflatable technologies with the goal of orbiting commercial space stations and establishing lunar bases. On the rare occasion he granted an interview, Mr. Bigelow rarely had anything good to say about NASA, accusing the agency of being an impediment to the commercial development of space. Perhaps NASA took Bigelow's criticisms to heart because, in 2004, NASA chief Sean O'Keefe told a presidential commission the agency would have to undergo a complete transformation if Americans were ever going to achieve a permanent presence in space. Significantly, he noted that a key part of the new strategy would be a reliance on private companies to do much of the work. It took a while for that transformation to take effect, but the launches of SpaceX's Falcon rocket underscored the fact that the most exciting developments in aerospace were no longer taking place at NASA, but in the commercial arena—thanks to significant funding from the agency it should be noted.

But was it a transformation? In 2014, you'd have to answer in the negative. While innovations in commercial space have the potential to dwarf those offered by current NASA programs, that potential is being stifled by funding and Congress. While SpaceX sends cargo supply vessels to the International Space Station (ISS) and plans manned Mars missions, NASA's current bold endeavour is plotting a trip to an asteroid sometime in the 2030s. What happened? Well, it certainly isn't NASA's fault. As a space agency, it remains probably the greatest assembly of talent in the US; lack of vision at NASA has never been the fault of its scientists or engineers, who routinely dream up everything from pioneering Mars mission architectures to interstellar probes. The problem is that many of these multibillion-dollar ventures begin in one administration, only to be canceled in the next. By Congress. Who else? It's a system of patronage that sees congressional partisans who couldn't care less about the space program placed in charge of it solely because of the presence of aerospace facilities in their districts. Not convinced? Take the Space (Senate) Launch System (SLS) as an example of where America's space program has derailed. The SLS (Figure 7.2) is essentially a large replica of the Saturn V and its lack of innovation means it will do very little to reduce launch costs or make space more accessible.

7.2   Space (Senate) Launch System. Courtesy: NASA

Initially priced at US$18 billion, this behemoth was designed for cargo delivery and exploration beyond low Earth orbit (LEO), with a project-completion date of 2017. As is so often the case with these projects, the costs have already been pushed to the right, with some estimates predicting costs having surged to US$22 billion. But what makes the SLS boondoggle so exasperating is not just how much is being expended for so little; it is imagining what could be done with those resources in the commercial spaceflight arena. Actually, you don't have to stretch your imagination that much: you just have to take a look at what Elon Musk has achieved with his Falcon 9 launch vehicle, which went from drawing board to completion in four years for just US$300 million. Now that's the power of the free market. And SpaceX isn't alone. Companies such as Blue Origin compete with SpaceX in pursuit of more affordable launch options, while other firms such as Bigelow Aerospace seek to accelerate the development of human space habitats. Each of these companies is charging ahead, armed with innovative plans, achieving dramatically reduced costs and doing it with enormous ambition. Each has the potential to be transformational, and each has already changed the commercial space landscape. Now imagine what could be done if resources being thrown at the SLS were repurposed for technology incubation and commercial project missions. For the cost of SLS, you could afford dozens of BA-330s. Dozens!

But Congress isn't the only one to blame. When President Obama was elected, NASA was busy working on the Constellation Program, which would have seen a return to the Moon as a first step towards the manned exploration of the Solar System. Lauded in the scientific and space community, and endorsed by Neil Armstrong, this highly ambitious project was canceled by the Obama Administration in the name of cost-cutting, only to be replaced by … the SLS. As Bigelow has argued so forcefully at so many conferences, it's high time to accelerate the development of the high frontier, but we have to act. Which is what Bigelow and NASA did, by burying the hatchet, engaging in partnership, and signing three Space Act Agreements (SAAs), which provide for an ongoing exchange of personnel and technology, the joint testing of Bigelow projects at NASA facilities, and the transfer of NASA patents to Bigelow. And perhaps the most visible result of this collaboration will be Bigelow's BA-330 module.

BA-330 AND BEYOND

October 2010, Las Cruces, New Mexico. A long time ago, Las Cruces was a Western frontier settlement and, if you happen to visit the city's Farm and Ranch Heritage Museum, you will see 19th-century pioneers' personal effects, farming tools, and wagons once drawn by horses. But, for the past several years, the museum has served as the site of the International Symposium for Personal and Commercial Spaceflight (ISPCS), a conference that deals with an entirely different frontier, albeit only 100 kilometers above the museum. The 2010 edition was the biggest yet, attended by 400 professionals from the commercial spaceflight arena—evidence of the growing interest in the civilian spaceflight business. Perhaps the main draw of the 2010 event was the attendance of Mr. Bigelow, who announced the biggest news of the symposium, a hint of which was displayed in the company's marketing booth that featured models of its space modules (Figure 7.3).

7.3 Model of the Bigelow inflatable. Courtesy: Paul Lemon

Table 7.1. BA-330 specifications.

Length	9.5 meters
Diameter	6.7 meters
Mass	20 tonnes
Crew	Six
Pressurized volume	330 cubic meters
Radiation protection	Equivalent of better than the ISS
Ballistic protection	Micrometeorite and Orbital Debris Shield provides superior protection to that of aluminum can designs
Electrical power	Independent system comprising solar arrays and batteries
Propulsion	Fore and aft of spacecraft; aft system can be refueled and reused
Avionics	Independent avionics system to support navigation, re-boost, docking, and maneuvering
Environmental control and life-support system (ECLSS)	Each module will contain its own ECLSS, including lavatory and hygiene facilities
Windows	Four, each coated with film for ultraviolet (UV) protection

The news was that six sovereign clients had signed memoranda of understanding to utilize Bigelow Aerospace orbital facilities: the UK, The Netherlands, Australia, Singapore, Japan, and Sweden.

There was considerable attention paid to models and rendered posters of facilities that utilized various configurations of clusters of Bigelow's workhorse, the BA-330 module (Table 7.1 and Figure 7.4). An impressive structure measuring almost 10 meters in length,

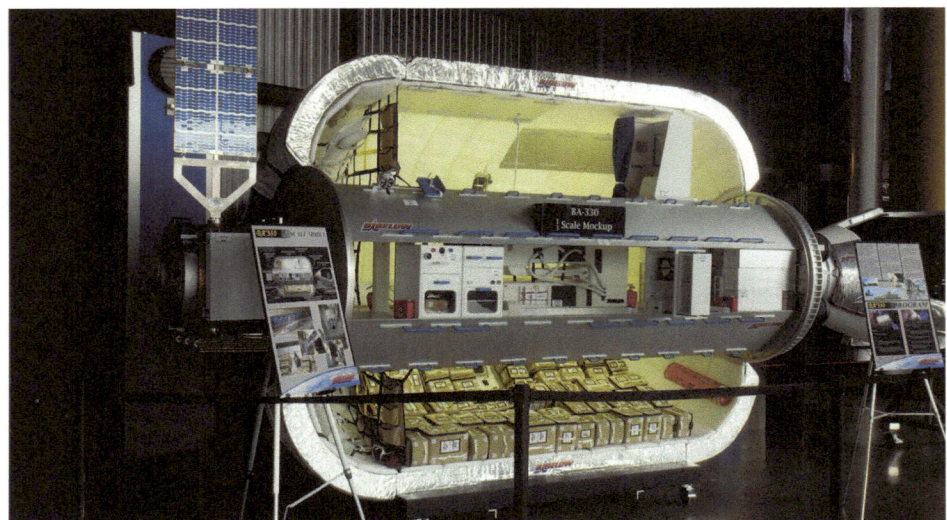

7.4   Model of the BA-330. Courtesy: Bill Ingalls/NASA

Table 7.2.   BA-2100 specifications.

Length	17.8 meters
Diameter	12.6 meters
Mass	70 tonnes
Crew	16
Pressurized volume	2,100 cubic meters

6.7 meters in diameter, and containing 330 cubic meters of volume, this module is capable of comfortably supporting six crewmembers, and is more than 10 times as large as the Orion configuration NASA has devised for its asteroid mission. But, large as the BA-330 is, even this capacious habitat is dwarfed by the monster module: the BA-2100 (Table 7.2), which is six times as large. Featuring multiple decks, the BA-2100's dry mass could be as low as 70 tonnes, which means that, in its un-inflated state, it could be placed into orbit by the SLS.

More details about the possible uses of these inflatables were published in two reports issued to NASA (but not made public) in 2013 under an unfunded SAA known as the Gate 1 Report in which Bigelow indicated he believes that "without cost effective human habitats, America's human space exploration ambitions throughout our solar system isn't going to happen". In the document, Bigelow proposes utilizing his BA-330, his Olympus modules, and a family of space tugs as examples of cost-effective habitats and transportation systems that can be used by NASA for exploring the Solar System. The document also provides a description and a price list of the habitats that he can offer the agency. Perhaps the most versatile is the BA-330, which can be reconfigured for just about any destination or purpose. The basic BA-330 model (Figure 7.5) is designed to serve as a LEO habitat,

7.5 The BA-330. Courtesy: Boeing

and can sustain a crew of six. Around the perimeter of the BA-330's airlock are two propulsive thrusters. The aft propulsion module system uses monopropellant hydrazine, while the forward propulsion system is a regulated gaseous hydrogen/gaseous oxygen system that is refillable through the environmental control and life-support system (ECLSS). The habitat also features four arrays which extend from the forward airlock, with two electricity-producing photo-voltaic arrays and two radiators, and will use NASA docking systems to support visiting vehicles, and to link with other modules as depicted in the diagram. The Gate 1 Report goes on to explain how the module, which has a design lifetime of at least 20 years, will most likely have a zero-G toilet with solid and liquid waste collection, semi-private berths for each crewmember, exercise equipment, a food storage and preparation station, and a personal hygiene station. For commercial customers, the

Table 7.3. Space tugs.

Tug	Features	Uses
Standard Transit Tug	Docking adapters fore and aft	Propulsion of habitat
Solar Generator Tug	Docking adapters fore and aft	Transport solar panels to habitats
Docking Node Transporter	Five port docking node/propulsion elements	Transport of docking node
Capture Tug	Grapple arms with multiple joints	Capture disabled spacecraft, satellites

BA-330 will cost US$25 million to rent a third of the station for 60 days. That price is in addition to the price of a seat on board SpaceX's Dragon, which will cost US$26.5 million. These prices will include consumables, on-board research equipment, a full-time Bigelow Aerospace crew (one pilot and another tasked with maintenance duties), and astronaut training.

Beyond a single BA-330 is the idea of joining two together to form a service station. The report suggests these platforms could be used by NASA to test hardware that can't be tested on the ISS. These service stations could also be reconfigured to serve as medical facilities and perhaps as an overhaul depot for the space tugs that will be zipping around LEO. And, on the subject of space tugs, Bigelow has a plan for those as well. The report suggests using a family of tugs to service Bigelow habitats: a Standard Transit Tug, a Solar Generator Tug, a Docking Node Transporter, and a Spacecraft Capture Tug (Table 7.3). Designed to be grouped together in various configurations depending on mission requirements, each tug is sized for launch on SpaceX's Falcon Heavy Rocket.

The next iteration of the BA-330 is the BA-330-DS, which is the deep-space version of the habitat, designed for use beyond LEO. The BA-330-DS will be very similar to its LEO cousin except for increased radiation shielding. Whereas the BA-330 only provides water tiles for sleeping quarters, the BA-330-DS will be outfitted with a full complement of water tiles that surround the interior. The module is designed to be launched on either a Falcon Heavy or an Atlas 552, before being tugged to its intended destination. Beyond the BA-330-DS is the BA-330-MDS (Modified Deep Space) for use on the Moon or perhaps Mars. Configured with landing propulsion buses for landing on a planet or moon, the MDS is the habitat Bigelow has identified as forming the core element of his lunar base, which is discussed in the following chapter.

Bigger than any of the BA-330 iterations is the whopping Olympus model, which offers 2,258 cubic meters of internal volume and can accommodate a crew of 24–30. It features a multiple-deck concept which means it can be configured for the needs of all sorts of programs and has a minimum design lifetime of 20 years. Although the Olympus module could be a payload for the SLS, it could possibly be launched on a Falcon Heavy.

Also under consideration are other inflatable permutations (Table 7.4), such as resupply depots, deep-space complexes, and medical facilities. And it isn't just Bigelow that is dreaming up these space facilities. Take Uptown (Figure 7.6), for example. Suggested by Marotta Space Research, Uptown is more a speculative exercise designed to get people thinking about how the next crop of space stations might evolve. With a ring 148 meters in diameter, this wheel in the sky would rotate at two revolutions per minute (15.5 meters per second), generating one-third of Earth's gravity. With an internal pressurized volume of 18,360 cubic meters, the station could accommodate 100 people in 11 residence quarters,

Table 7.4. Possible configurations.

Space Complex Bravo	Resupply Depot Hercules
1,320 cubic meters	8,300 cubic meters
4 BA-330 modules	6 BA-330 modules
2 Large propulsion buses each with docking	3 BA-2100 modules
nodes	3 Large propulsion buses each with docking
2 Crew capsules	nodes
	3 Crew capsules

Deep Space Complex	Advanced Medical Facility
1,320 cubic meters	3,000 cubic meters
4 BA-330 modules	9 BA-330 modules
9 Large propulsion buses	3 Large propulsion buses
3 Docking nodes	3 Crew capsules
Crew capsule	

four quarters with a capacity for four tourists each, and seven non-tourist quarters with a capacity for 12 non-tourists each. Based on a vacant mass (just the mass of the station without fittings) of 2,548,000 kilograms, and assuming a launch cost of US$2,547 for the proposed cost-per-kilogram on the Falcon Heavy, the station's cost would be US$6,489,756,000 to launch to LEO.

In Figure 7.6, BA-330 modules (28 of them) are depicted in green and the (pressurized) corridor modules in pink. The corridor modules measure 19 meters in length, five meters in width, and feature a foyer nearly seven meters long and two meters wide. The blue elements are unpressurized spine trusses (24 of them), each 18 meters long. Corridor segments are connected to the spine in addition to the support trusses for power plants and radiators. In yellow are the 88 solar power plants, each generating 25 kilowatts of baseline power for a total of 2,200 kilowatts while the station is facing the Sun. This energy will charge batteries distributed throughout the corridor for use when the station is transiting the night-side of Earth. The radiators (brown), each 125 square meters in size, can dissipate 149 kilowatts of energy for a total dissipation capacity of 13,140 kilowatts. Uptown is a unique concept that isn't feasible to construct in the short term, but nevertheless represents an interesting example of the out-of-the-box thinking that one often associates with NASA's Institute for Advanced Concepts. But, before even one BA-330 is orbited, sovereign customers will want to know that all systems have been tested, so let's turn our attention to the life-support system.

LIFE SUPPORT

One of the key elements of Bigelow's space stations will be the life-support system. For several years, Bigelow has partnered with Madison, Wisconsin-based Orbital Technologies Corporation (ORBITEC), whose Human Support Systems and Instrumentation Division

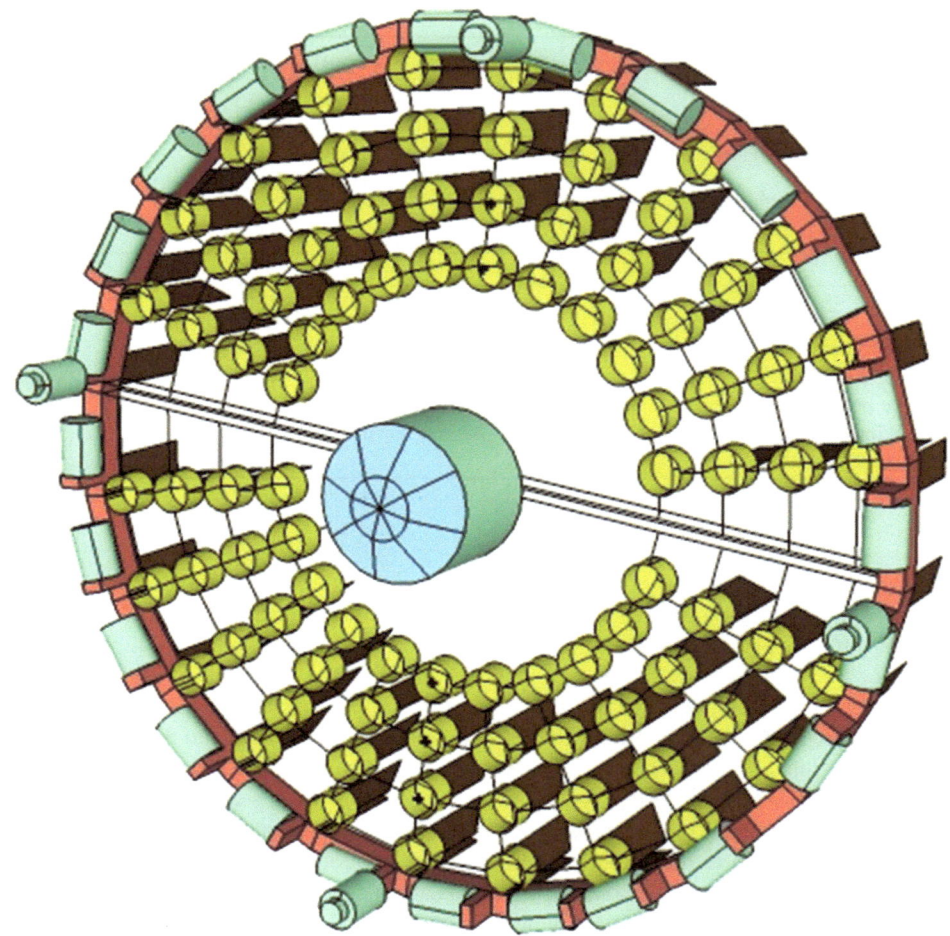

7.6 Uptown's vision of how inflatable space stations might evolve. Courtesy: Marotta Space Research

has worked with Bigelow to develop an ECLSS. With more than 20 years' experience developing closed human environments for space travel, ORBITEC was a logical choice to develop the myriad life-sustaining systems required for Bigelow's family of orbiting outposts. What follows is a short description of the primary life-support systems on board the BA-330.

A major concern during manned spaceflight is the quality of the cabin atmosphere. Ever since there have been astronauts, engineers have sought to prevent the uncontrolled accumulation of gaseous trace contaminants arising in the confines of the spacecraft. These contaminants could arise from off-gassing (from structural materials, electronic equipment, or materials used in experiments), from system failures (leaks or equipment over-heating), or from metabolic products produced by the crew. To ensure a safe atmosphere on board the BA-330, these trace gases will need to be monitored and controlled so they remain below safe limits. In the world of manned spaceflight, these limits are known as the Spacecraft Maximum Allowable Concentration (SMAC) values and have been defined as a result of medical considerations, previous spaceflight experience, and analogous terrestrial experiences such as in submarines. SMAC values vary considerably according to the type of chemical compound and also as a function of the mission's length (time of exposure), with higher SMACs generally being acceptable during shorter missions. The monitoring of trace gases is limited only to the most toxic or explosive ones such as carbon monoxide and hydrogen, while the atmospheric constituents of nitrogen, oxygen, water vapor, and carbon dioxide are monitored by dedicated sensors.

Another function of the BA-330's life-support system will be to supply oxygen. Most spacecraft carry their own supply of oxygen with them, but the missions of these spacecraft generally last a short time, in the order of days to two weeks. In contrast, the BA-330 is designed for long-term spaceflight, so it will need to use oxygen generators and pressurized oxygen tanks. These oxygen generators will make oxygen from water by a process called electrolysis, during which an electric current passes through water from one positively charged electrode called an anode to a negatively charged electrode called a cathode. In addition to supplying oxygen, Bigelow's customers will also need a steady supply of water, which will be limited, just like it is on the ISS. If you happen to be one of the lucky sovereign astronauts flying on the BA-330, you will be washing your hands with less than one-tenth the water that people typically use on Earth and, instead of using 50 liters to take a shower, which is typical on Earth, BA-330 occupants will use less than four liters. Even with intense conservation and recycling efforts, the BA-330 will gradually lose water because of inefficiencies in the life-support system. Water will be lost by the BA-330 in several ways: the water recycling systems will produce a small amount of unusable brine, the oxygen-generating system will consume water, air lost in the air locks will take humidity with it, and the carbon dioxide removal system will leach water out of the air. This means water will need to be resupplied, because none of the current water reprocessing technologies are 100 % efficient.

Temperature control will be another important element of the life-support system. This is not only to ensure the comfort of the astronauts, but because most active equipment can only work at room temperature. The solution to the temperature control problem is a good thermal design to protect the equipment from hot and cold temperatures, either by proper heat or cold insulation from external sources, or by proper heat or cold removal from internal sources. To ensure humans and equipment continue to function optimally, the BA-330 will use an active thermal control system using a single fluid loop that utilizes a human-safe fluid.

While the inflatable module concept has been subject to rugged leak tests, the BA-330 will still need to find room for air supplies for spacecraft re-pressurization and crew respiration. Ambient cabin pressure will need to be maintained, and atmospheric leakage countered. This will be achieved by periodic injection of gas, the amount calculated from measurements of total cabin pressure and the partial pressure of oxygen. For example, if the BA-330's internal pressure rises too much, a controlled release of some of the atmosphere will be necessary before the ambient pressure becomes unpleasant for the occupants: in the case of most spacecraft, this would also be a hazard because of the risk of exceeding the structural limits of the spacecraft, but these inflatables have been tested to twice that required by ISS modules, so no concerns there. In multi-module spacecraft such as the ISS, pressure gradients between or among modules are rectified through pressure equalization valves between compartments, and ground control and crew can control these valves and regulate ambient pressure. Although the BA-330 is a single module, it is probable that crews will utilize a similar system.

Another function of the life-support system will be carbon dioxide removal. In the atmosphere, carbon dioxide concentrations are approximately 0.04% but, on board spacecraft, the carbon dioxide concentration can get much higher, which poses a problem because carbon dioxide is toxic. As carbon dioxide concentration increases, astronauts suffer certain symptoms:

- at 1%: drowsiness
- at 3%: impaired hearing, increased heart rate and blood pressure, stupor
- at 5%: shortness of breath, headache, dizziness, confusion
- at 8%: unconsciousness, muscle tremors, sweating
- above 8%: death.

On board the ISS, in the US Destiny lab and Node 3, a carbon dioxide removal assembly uses molecular sieve technology to remove carbon dioxide. The molecular sieves are zeolites—crystals of silicon dioxide and aluminum dioxide, which arrange themselves to form tiny screens. The openings of the screens or pores are consistent sizes that allow some molecules to enter and get trapped in the sieves. In the removal assembly, there are four beds of two different zeolites: one type that absorbs water and another type that absorbs carbon dioxide.

In March 2011, Bigelow completed a closed-loop test of a prototype ECLSS inside the company's newly constructed test chamber. The test locked three Bigelow engineers inside the 180-cubic-meter structure for eight hours, during which they performed a variety of tasks that demonstrated the system's ability to control temperature, humidity, pressure, oxygen content, and the removal of carbon dioxide and trace-gas contaminants from the environment. The test chamber was designed to replicate the interior volume and shape of the three-person *Sundancer* module, although the facility is scalable to the larger BA-330 module. The life-support system and its test chamber were built in-house, giving the company more control over system development and allowing fast manipulations to the design to counter issues or meet efficiency gain. In addition to completing the demonstration facility, Bigelow Aerospace also wrapped up work on an analytical chemistry laboratory that allowed real-time monitoring and analysis of gases or liquids in the test chamber's atmosphere. Bigelow is simply taking NASA's technology development and packaging them into a more producible form. While it may not be cutting-edge in terms of technology, it is definitely cutting-edge in terms of affordability and availability.

PROPULSION AND AVIONICS

The ISS requires an average of 7,000 kilograms of propellant each year for altitude maintenance, debris avoidance, and attitude control. Since Bigelow's modules will also need to dodge space junk and boost their orbits, the modules will feature a propulsion system, although, according to the Gate I Report, those space tugs mentioned earlier may also be employed. With this in mind, Bigelow Aerospace inked a multi-million-dollar contract with Orion Propulsion Inc. of Huntsville, Alabama, to supply the attitude control system for the forward end of its *Sundancer* module. Bigelow also awarded a US$23 million deal to Aerojet-General Corporation of Sacramento, to supply the propulsion system for the aft end of *Sundancer*, as well as a system to handle rendezvous and docking. While *Sundancer* has been removed from Bigelow's station evolution path, the BA-330 will likely employ similar systems.

The avionics flight hardware was built by Andrews Space, Inc., under a five-month fixed-price contract. Andrews developed, flight-qualified, and delivered 20 propulsion system diode boards responsible for electromotive force protection within the Aerojet propulsion system. Under a second contract, Andrews was tasked with developing fault-tolerant propulsion controllers which would have interfaced with the Bigelow *Sundancer* human-tended space platform and control individual propulsion system valves and heaters based on commands received over an Ethernet communication bus—again, since the demise of *Sundancer*, it is likely similar systems will be used on the BA-330. The controller also performs fault-detection, isolation, and recovery based on propulsion system instrumentation.

PURPOSES

All sorts of suggestions for commercial space stations have been made over the years. Bigelow's suggestions include using the modules as medical facilities, resupply depots, and deep-space exploration habitats. They could be used as manufacturing facilities for products that can only be made in microgravity or they could be transit ports for spacecraft heading beyond Earth orbit. They could be used as test platforms, as research facilities, or as a biological containment facility. How they will be used is still the subject of debate, but what we do know is that, in 2014, commercial cargo and crew only have the ISS as a destination. And, when the ISS is de-orbited and splashes into the Pacific sometime in 2024, there will be no destination, which should open up a market for one or more of Bigelow's commercial stations. What kind of station will be the most commercially successful? It's difficult to say, but commercial crew and cargo will need thriving and growing destinations to have any significant growth potential and to make a return on investment. Space tourism would be greatly helped with a destination that has room for people to move and float around in, but that's just one part of the equation: this market isn't going to grow unless launch costs decline. Even Bigelow's costs (US$26.25 million a seat on board a Dragon) are prohibitive for all except a wealthy few, but research and tourism aren't the only viable businesses with LEO potential. Take satellite servicing, for instance. For this to emerge as a business, it could utilize on-orbit storage of spare parts and fuel, which could be transported by those space tugs mentioned earlier. And those tugs would need an operating base and a fuel depot to succeed.

BIGELOW'S BUSINESS CASE: IT'S NOT ABOUT SPACE HOTELS!

So, if you happen to be a potential sovereign customer, what will you get for your US$51.25 million (US$26.25 million for the launch on board Dragon plus US$25 million for your 60-day stay on board a BA-330)? Well, that price includes the costs of the launch and range fees and cost of recovery in addition to the cost of training the client—a subject we'll get to shortly. Liability insurance will be separate from whatever the launch provider may or may not have and Bigelow will provide its own insurance to its customers. But will this business of launching customers to inflatable habitats be profitable? That's hard to say because there are so many variables. For one thing, launch costs will be a function of traffic and the frequency and number of flights. Launch costs can also be a function of the ability of the launch provider to become more cost-efficient in the fabrication of their vehicle (the launch vehicle and/or the capsule). In setting out his business plan, Mr. Bigelow is always keen to emphasize that his concept of space outpost utilization is not about space hotels, despite what the media say. Orbiting Bigelow habitats has *never* been about space hotels and you get the sense when talking to him that he's become more than a little frustrated with media reporting that tars his space habitats as space hotels just because, on Earth, he happens to be president of the Budget Suites chain. It doesn't seem to matter how many times Mr. Bigelow explains his business case, the media seem determined to portray Bigelow Aerospace as the space hotel folks. So, in case any media people happen to read this book, the business of BA-330s being space hotels is not the case and it never has been the case. Bigelow has always imagined building commercial space habitats that can accommodate a variety of functions: from governments aiming to pursue their own space programs to researchers wanting to conduct science. Not to take anything away from space tourism but, to create a thriving private sector space industry, there will need to be more than tourism. Ultimately, Bigelow feels that it will be applications derived from space research that will drive the engine of space development. Bigelow also reckons the innovations that will support such an industry have yet to be discovered, which is why Bigelow Aerospace has such a strong interest in space researchers as a customer base.

When commercial real estate developers build an office building, they usually don't have to pre-sell all the space before they start construction. Instead, the company usually signs an anchor tenant to secure financing. Could such a model work for commercial space stations? Obviously it would help to have government and commercial tenants signed to get financing for a commercial station, which is why Bigelow has those agreements in place with more than half a dozen sovereign nations. But will they take the risk? Well, no one really knows whether revenue can be guaranteed but, once it does happen, this commercial space station business suddenly becomes more interesting to the financial community. The question then becomes: who will the anchor tenant be? The answer is the government, simply because no one else has pockets deep enough. Whether that happens is in the hands of the politicians and NASA's long-term vision, which is something Congress can't agree on. NASA wants to move beyond LEO and turning LEO over to potentially more cost-effective business models such as orbiting BA-330s and BA-2100s would free up resources to develop payloads for exploration-class missions. So, if the President and Congress ever agree on a vision for NASA, even in this fiscally restrained environment there could be significant human exploration missions not far in the future.

As for that Gate Report, Bigelow has stated that his company has the financial wherewithal to pay for at least two BA-330s habitats which should be ready by the end of 2016 and they also have the capacity to pay for two transit tugs which could be used in conjunction with their habitats (one of the tugs would have a docking node on it and could use a Bigelow Expandable Activity Module (BEAM) for extravehicular activities (EVAs). But Bigelow cautioned that his company could not fund the entire operation without some commercial funding and they need to sign enough customers to pay for half the cost of launching the modules into LEO. Ultimately, Bigelow acknowledged the commercial sector will need NASA as an anchor tenant to have a business case for exploration beyond LEO. The reason? The commercial space industry is just findings its legs and, without an anchor tenant, it just doesn't have the strength to execute deep-space missions. In short, those lunar bases that are discussed in the next chapter may have to wait.

In 2014, Bigelow's BA-330 is essentially ready to launch at any time: it's just a question of demand and a launch vehicle. As we know, the launch vehicle question is down to Congress and funding being available. Demand? Well that depends on the business case. Which foreign and domestic companies could benefit from utilizing expandable, orbital space habitats and what are the profit margins for Bigelow? Well, let's crunch some numbers (Table 7.5). If building a BA-330 costs US$125 million and launching it on a Falcon Heavy costs another US$80 million, then each module will have a capital cost of US$205 million. Now, if each of the three sections is rented for US$25 million for 60 days, the maximum revenue for a BA-330 module would be about US$450 million per year. Let's assume three cargo flights are needed to resupply the module at about US$100 million per flight and operation support costs another US$50 million per module per year. That brings

Table 7.5. Projected revenue for Bigelow's BA-330s.

Capital costs	
A. Cost of building BA-330	US$125 million
B. Cost of launching BA-330 (via Falcon Heavy)	US$80 million
C. Capital cost (A + B)	US$205 million
Operating costs	
D. Three cargo resupply flights/US$100 million per flight	US$300 million
E. Operation support costs	US$50 million
F. Annual costs (D + E)	US$350 million
Revenue	
Monthly revenue (assuming maximum occupancy)	
G. 3 × US$25 million	US$75 million
Annual revenue	
H. (G × 6)	US$450 million
Return on investment (assuming 10-year lifespan)	
I. Annual operating costs (10% of C + F)	US$320.5 million
J. 10-year annual revenue (H × 10)	US$4.5 billion
K. 10-year profit (J − 10 × I)	US$1.295 billion

the annual costs to US$350 million. If the module has a 10-year lifespan (Bigelow reckons the modules could last 20 years, but let's be conservative here), then the total available after paying off the capital investment for profit is ~US$900 million per module on just a US$205 million investment. That's a healthy return!

These calculations assume an occupancy rate of 100%, but how likely is that scenario? Truth is, we just don't know but, even with a reduced occupancy rate, the numbers look good. US$51.25 million for 60 days (the cost of the flight and the occupancy) isn't that much money when you consider that spaceflight participants are paying US$35 million[1] or more for just 10 days on board the ISS. And don't forget that these spaceflight participants must spend at least six months in Star City training for the mission. They also have to learn Russian. That length of time training will no doubt prove costly to most of the people with the financial wherewithal to afford such a flight in the first place. It was a subject I discussed with Mr. Bigelow in 2005. He was of the opinion that most people with that sort of money wouldn't be prepared to spend the best part of half a year training for a flight, especially if that training required them to be away from family and friends. I agreed, and we discussed how long the training had to be. I made a few notes on one of Marie Callender's napkins and came up with a five-week training plan. Offer a five-week training plan to potential customers, dangle the added carrot that they don't have to learn Russian, and chances are demand will go up. Make it even more attractive by not having your customers wait years for their flight opportunity, make the whole process akin to booking an airline ticket, and demand will go even higher. Also, remember that these numbers are estimates.

According to Bigelow, if the company attracts enough customers to lease his inflatable modules, it could mean up to 25 launches a year, which would mean good business for Cape Canaveral and Space Florida, which continues to pursue a number of strategies to drive growth of the commercial space industry in the Sunshine State. Since Bigelow's ambition is to launch and operate as many modules as possible, it makes good business sense that Space Florida wants to develop a partnership with the company. The goal of such a partnership would not only be to establish a significant Bigelow presence in Florida, but also to leverage both company's relationships to attract new, internationally based customers. Such a partnership will no doubt help build a dynamic future for on-orbit commercial space operations, increase launch activity, and grow the aerospace field in Florida.

LOOKING AFTER THE CUSTOMERS

In 2010, Bigelow advertised for astronauts (see sidebar) to work in Mission Control, to develop training programs, and to work on board the Bigelow Aerospace Station Complex. In 2010, employing astronauts to train sovereign clients made sense because there were no companies offering commercial astronaut training but, in 2014, that's changed.

---

[1]   Sarah Brightman was rumored to have paid US$50 million for her ticket to the ISS. The soon-to-be sopranonaut will visit the station in 2015.

Bigelow Aerospace seeks professional astronauts* to fill permanent positions. Qualified applicants need to have completed a training program from their government or recognized space agency and have at least some flight experience on a recognized space mission. Specialized training and/or experience (i.e. Medical, Payload Specialist, EVA, Pilot, etc.) is not a pre-requisite, but is definitely a plus.

---

* In July 2014 Bigelow Aerospace announced it had hired former NASA astronauts Kenneth Ham and George Zamka to form the cornerstone of the private astronaut corps. Zamka came from the Federal Aviation Administration's Office of Commercial Space Transportation, where he was deputy associate administrator from March 2013. Ham, a Navy captain, had been chairman of the Aerospace Engineering Department at the U.S. Naval Academy in Annapolis, Maryland. He will begin developing a training program for the astronauts Bigelow hopes to recruit, and start work on operational protocols for Bigelow's habitats.

Possible opportunities will come in two areas:

1. Ground

   - Working with Marketing Team to secure government and corporate clients
   - Working with Design and Fabrication Teams to help optimize layout of systems for on-orbit serviceability and ergonomics
   - Working with Mission Control Team on final checkout of flight vehicles, both pre and post launch
   - Help Develop Astronaut training programs for Bigelow Aerospace Professional Astronaut Corps as well as Client Astronaut Corps
   - Work instructing in the Bigelow Astronaut Training Program

2. Flight

   - Perform as Professional Astronaut aboard Bigelow Aerospace Station Complex
   - Manage all on-board aspects of employee and customer astronaut personal safety
   - Maintain the Station Complex as required (mainly IVA, but some EVA as well)
   - Help clients with payloads or experiments (primarily with regards to integration into station's systems and communications)

WAYPOINT 2 SPACE

(www.waypoint2space.com)

Have you ever wanted to be an ASTRONAUT?

To go into space, step out of the vehicle, and float above the Earth while reaching for the stars—but wondered if you have what it takes? For the first time in history, you can train like an astronaut using the most advanced facilities and equipment in the world. Headquartered at NASA's Johnson Space Center, we offer the definitive training experience with our fully comprehensive and immersive space training programs. These one-of-a-kind programs prepare you for spaceflight while you experience firsthand what every astronaut has during their preparation for space. Additionally, SFP's are trained in accordance to our FAA Safety Approval ensuring a consistent level of spaceflight competency.

Want to train like an Astronaut? Then Waypoint 2 Space is your first step.

Click on the Waypoint 2 Space website and you'll see the above introduction. Based in Building 35 on the NASA Johnson Space Center (JSC) campus in Houston, Texas, Waypoint has the cachet of being affiliated with NASA, which may be the reason for the prices of its training programs. The company offers a basic spaceflight course, suborbital training programs, and orbital training for commercial payload specialists. If Bigelow were to use Waypoint, his clients would be enrolled on the Orbital Training program, which consists of eight weeks of training, although this expands to 12 weeks with EVA training. Students enrolled on this program (the first course is due to begin in 2015) can look forward to instruction on spatial disorientation, emergency depressurization procedures, vehicle malfunctions, and contingency operations. The program also covers topics such as launch and orbital vehicle operations, and intra-vehicular activity (IVA). At the end of the program (Appendix II), the newly minted spaceflight participant will:

- be competent in orbital operations, vehicle systems, environments, and physiological effects associated with extended duration spaceflight
- acknowledge and self-correct physiological issues associated with spaceflight
- be able to operate in confined spaces during extended spaceflight durations
- be competent in operations requiring use of a partial pressure suit
- exhibit adaptation to and operate in microgravity for extended durations
- be competent in nominal and off-nominal vehicle conditions during orbital flight
- demonstrate ability to work within a crew environment under stressful situations.

The cost? A cool US$45,000, although this includes a cool flight suit and jacket. Will Bigelow send his clients to Waypoint? We don't know. But what are the options? Well, one (cheaper) alternative is the American Astronautics Institute.

AMERICAN ASTRONAUTICS INSTITUTE

(http://astronauticsinstitute.com)

Do you have what is takes to be an astronaut?

We graduate Scientist-Astronauts, not 'Spaceflight Participants'

The American Astronautics Institute provides a hands-on, practical education and astronaut training for the next-generation of commercial scientist-astronauts near NASA's Kennedy Space Center, FL.

That's the pitch on the American Astronautics Institute's (AAI) website. AAI is the education affiliate of Jason Reimuller's Integrated Spaceflight Services, founded in 2010 from a group of NASA scientists and engineers. The institute was established to help aspiring space scientists get their ideas funded and off the ground and also to train astronauts. To achieve that goal, AAI has recruited top scientists and engineers and developed their programs with a science emphasis, which are perfectly suited for Bigelow's clients. In common with Waypoint, AAI offers a variety of programs, including its flagship Flight Engineer Operations program (Appendix II), priced at a competitive US$34,500. A graduate-credential level program that integrates professional astronaut training, the Flight Engineer Operations course combines four months of virtual instruction with a 22-day intensive training program near NASA's Kennedy Space Center. It also provides a comprehensive education for those with an orbital flight on the horizon, training students to manage space missions, to operate spacecraft systems, prioritize mission objectives, and manage contingencies.

Those familiar with the timeline of government-sponsored astronaut training may be scratching their heads wondering how an astronaut can be trained in a matter of weeks, when professional astronauts take more than a year. The answer is simple. Astronauts on board the ISS dedicate the lion's share of their time to supporting station operations and maintenance, whereas astronauts on board Bigelow's Alpha Station will be able to focus exclusively on their own experiments and activities. The reason Bigelow's clients won't be distracted by plumbing issues or servicing troublesome equipment is because there will be two astronauts employed to do just that (see advert). Today's ISS six-person crews are lucky if they log 40 hours of research a week because there is just so much housekeeping to attend to. Not so on Bigelow's station. And, while no country has signed a contract and put a deposit down for an orbital stay, chances are that when they do, they could be trained by one of the companies described here.

8

The Man Who Sold the Moon

"But I don't think we'll go there until we go back to the Moon and develop a technology base for living and working and transporting ourselves through space."

Jack Schmitt

In *The Man Who Sold the Moon*, Robert A. Heinlein's classic science-fiction novella, Delos D. Harriman is a businessman determined to control the Moon. Most dismiss Harriman's endeavor as foolhardy, costly, and almost certainly unprofitable, but Harriman is undeterred. He asks his business partner, George Strong, and other tycoons to invest in the venture and, to solve the tougher financial problems, Harriman exploits commercial and political rivalries. He convinces the *Moka-Coka* company, for example, that rival soft-drink maker 6+ plans to turn the Moon into a massive billboard, using a rocket to scatter black dust on the surface. To an anti-Communist associate, he suggests the Russians may print the hammer and sickle on the face of the Moon. To a television network, he offers the Moon as a reliable and uncensorable broadcasting location. Since Harriman wants to prevent government ownership of the Moon, he uses a legal principle that states property rights extend to infinity above a land parcel on Earth. Because the Moon passes directly overhead only in a narrow band north and south of the equator, Harriman persuades countries to assert their rights and convinces the United Nations to assign management of the Moon to his company. Harriman sells land and naming rights to craters, and starts rumors that diamonds exist in lunar dust, intending to secretly place gems in a rocket to convince people the rumors are true. Harriman wants to be on the first flight of the *Pioneer* but the ship only has room for one pilot, Leslie LeCroix. *Pioneer* lands on the Moon, and returns to Earth, where Harriman is the first to open the hatch and he asks LeCroix for the "lunar" diamonds. The pilot complies, then produces real lunar diamonds as well. As Harriman predicted, once the first flight succeeds, many seek to invest in his venture to make more flights. The next flight will begin a lunar colony. Harriman intends to be on the ship, but the majority owners of the venture object to his presence on the flight because he is too valuable to the company to risk in space.

© Springer International Publishing Switzerland 2015
E. Seedhouse, *Bigelow Aerospace: Colonizing Space One Module at a Time*,
Springer Praxis Books, DOI 10.1007/978-3-319-05197-0_8

Sixty-two years after Heinlein's classic was published, another businessman announced his plans for the Moon, calling for the Federal Aviation Administration (FAA) to allow property rights for lunar mining. It was an announcement space exploration advocates had been calling for since the days of the lunar landings. At a NASA press briefing in November 2013, Mr. Bigelow explained he wanted private space companies to take a larger role in expanding NASA's astronaut explorations beyond low Earth orbit (LEO) and that he wanted the US government to offer those firms rights to mine the Moon. He went on to explain that lunar bases should be anchored to permanent commercial facilities and that, without property rights, there would be no justification for investment and the risk to life. It was a fair point, but the FAA couldn't help Bigelow because the agency only regulates launches and re-entries of rockets from orbit, and doesn't oversee activities of spacecraft. The latest US Commercial Space Act wasn't much help either because it doesn't mention lunar mining. The Outer Space Treaty (signed by all the major space-faring nations) is a little more restrictive (see sidebar) because it prohibits nations from claiming territorial rights on the Moon, which is widely seen as precluding them from awarding property rights to lunar resources. It's a gray legal area that may take some time to resolve but, of all the science-fiction scenarios slowly creeping into reality, a Moon base (Figure 8.1) is high on the list. And there's never been a shortage of ideas of how to achieve this goal: from habitats built on Earth and launched to the Moon to a 3D printed base made from lunar materials. But perhaps Bigelow Aerospace has the best shot of getting into the Moon base game. Here's how he may do it.

The Outer Space Treaty

Not surprisingly, Mr. Bigelow is filing for an amendment to the 1967 Outer Space Treaty to allow private individuals to own sections of the Moon, arguing that multiple entities, groups, and individuals should have the opportunity to own the Moon. The Outer Space Treaty, first created by the UK, US, and Soviet governments in 1967, and administered by the United Nations, is the basis of today's space law and has 102 signatories. Article Two states:

"Outer space, including the Moon and other celestial bodies, is not subject to national appropriation by claim of sovereignty, by means of use or occupation, or by any other means."

But, Article Six covers non-governmental entities operating within their jurisdiction, and that's the part Bigelow wants to change.

8.1   Moon base. Courtesy: NASA

THE SURFACE ENDOSKELETAL INFLATABLE MODULE

NASA has been in the lunar base business before, when its Exploration Vision was up and running. In 2005, in response to the need for viable surface concepts to realize this vision, the design team at Synthesis International created a preliminary design of a Surface Endoskeletal Inflatable Module (SEIM) that adapted the lessons learned from TransHab to the fundamental element of the critical path for human exploration on the Moon.

The SEIM (Figure 8.2) is similar to TransHab, comprising a rigid core from which a structural membrane is inflated to provide an expanded habitable space. The rigid core, formed by eight longerons supporting two end-cones, remains fixed in the launch and deployed configurations. Flat modular panels are packed around the longerons to provide stiffness during transfer and are subsequently deployed to form floors and partitions. The structural shell is stowed around the core in the launch configuration and is activated only after deployment to contain the atmosphere. In common with the TransHab technology, the SEIM is highly adaptable, with versions of the concept capable of being configured for all sorts of functions including laboratory, pressurized storage, and habitation.

The 2005 study, which focused on the habitation version of the module, defined a preliminary interior layout, dividing the module into private and public areas. Based on this design, each crewmember's private suite comprised a sleeping/living area, a desk/work station, and a place for storage, while the public area comprised a galley, a dining/meeting area large enough to accommodate the whole crew, an exercise area, an entertainment area, bathing and toilet facilities, systems control and command, and a communications area. The pressurized volume of the SEIM was 280 cubic meters (the BA-330's volume is 330 cubic meters) excluding the end-cones, and the floor area was 111 square meters, of which 62 square meters had at least 2.15 meters of clear height. In short, the SEIM offered its occupants plenty of space, which in turn would result in higher crew productivity and higher crew morale. The SEIM would also offer its residents peace of mind thanks to the

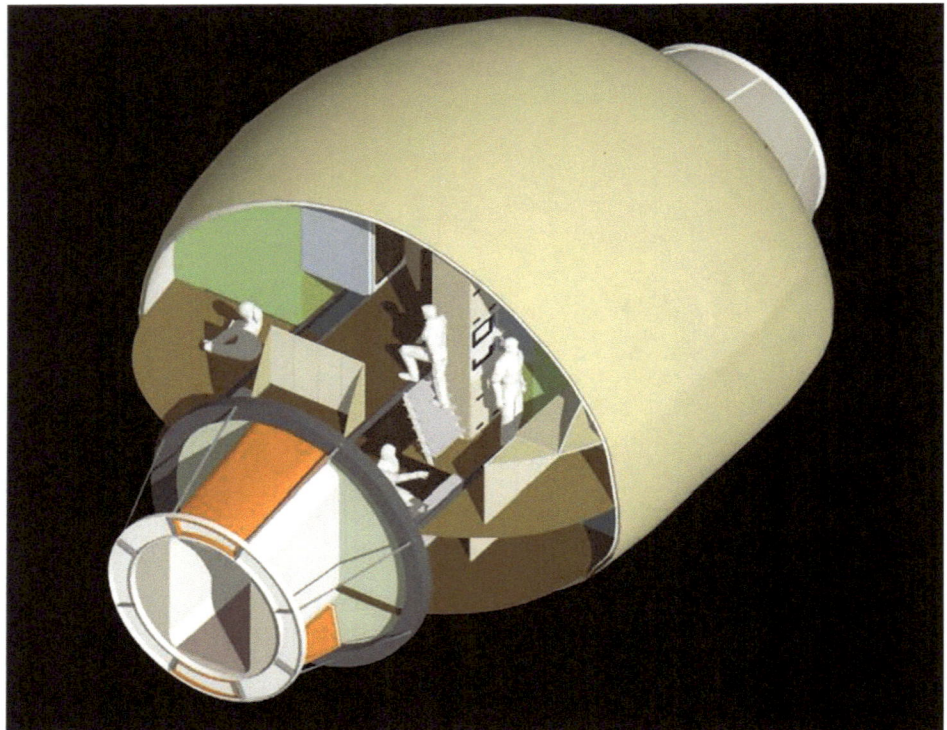

8.2   Surface Endoskeletal Inflatable Module. Courtesy: NASA

tough design: in common with the TransHab, the SEIM was designed so that even multiple hull breaches would not result in the structure's destruction since the shell was designed to absorb or deflect much of the energy from any impact. From a mission designer's perspective, the SEIM was a godsend because it maximized habitable volume while reducing launch mass, thereby simplifying mission architectures and reducing the number of launches needed. In short, the SEIM was perfect for NASA's vision for manned exploration: once again, inflatable structures were identified as a critical-path, major enabling technology—it was simple, it was safe, and it was cost-effective.

You may be wondering why NASA didn't just use a TransHab for its lunar plans, but this deployable structure was designed to be placed in orbit and the same design couldn't be used for a lunar base—not without some significant and radical adaptations, which is what the team set out to do. The end result was a structure (Figure 8.3) comprising a rigid core from which a structural membrane was inflated to provide a living space. The rigid core was formed by eight longerons supporting two end-cones, and flat modular panels were packed around the longerons to provide stiffness during launch and landing. Following landing, the longerons would be used as floors and partitions.

The SEIM frame, which is made from carbon-fiber-reinforced epoxy composites, has four components: end-cones, longerons, movable modular panels, and a static panel at the

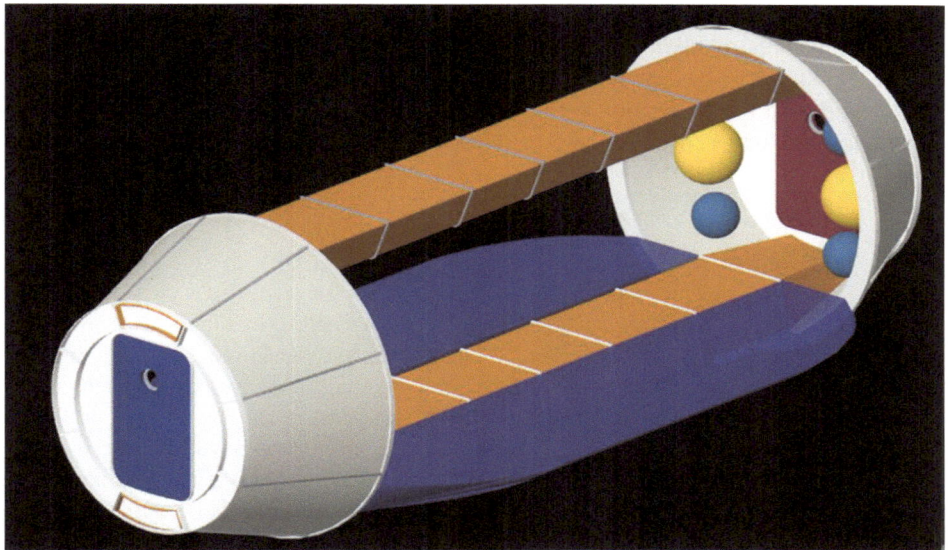

8.3   SEIM structure. Courtesy: NASA

bottom of the module. The purpose of the end-cones is to connect the inflatable shell and the frame, and also to provide a means to attach to the launch platform and serve as airlocks once the module is deployed. To resist the launch loads, the longerons are made of composites and provide the primary load-resisting elements of the frame. When the module is deployed, the longerons resist the tension load of the internal pressure.

From the inside out, the SEIM shell comprises several layers. An inner liner provides fire-retardant and abrasion protection, while three bladders form redundant air seals, with another four layers of felt providing evacuation between bladder layers. Internal pressure is resisted by woven straps of fiber material such as Kevlar that form the structural restraint layer. The module's transfer configuration is similar to the TransHab's, with the material being folded and compressed around the core. After landing, the module is inflated and it is here that the design deviates most from the TransHab. Since the SEIM's intended use is as a surface base, it has to have a floor, which has to be supported. This is achieved by the use of static longerons as horizontal beams in a configuration that results in a non-circular and non-radially symmetrical cross-section. This non-circular shape also required an upgrade from the TransHab design because the structural restraint layer was affected. In the circular TransHab, a constant stress gradient was easier to achieve thanks to the module's circular shape. With the SEIM, this was no longer the case because the bottom (floor) of the module meant the woven straps could no longer complete circles. To get around the problem, the SEIM engineers redesigned the weave pattern of the structural restraint layer to accommodate the new shape while keeping a constant stress gradient. Then they went to work increasing the robustness of the internal and secondary structures, which needed to be stronger because of the unsymmetrical shape of the surface module.

Because of funding challenges and changes in NASA's vision, the SEIM engineers never had the opportunity to further iterate, develop, and test the concept internally, which meant the SEIM was never flight-tested. But, for one private aerospace company, the SEIM work laid the foundation for what might eventually become the first commercial lunar base—as long as the Chinese (see sidebar) don't get there first!

Solar System Monopoly.

Despite an international treaty forbidding direct ownership by claim, use, or other means of the Moon by any one country or organization, the Chinese are planning to establish a lunar base sometime in the 2020s. If that happens, chances are China will be the one calling the shots, which is a scenario that makes Mr. Bigelow uncomfortable. "This will characterize the 21st and 22nd centuries and beyond. If we ignore this, it will be at our extreme peril," Bigelow told his audience at the International Symposium for Personal and Commercial Spaceflight in 2011. Is he right? Even if he is, who's going to stop red flags being planted on the lunar surface? There may be objections raised by some of the space-faring nations (probably the same nations exploiting resources in Antarctica, but that's another story), but it's unlikely this will prevent the Chinese laying claim to mineral rights and setting up mining operations, extraction facilities, and maybe even manufacturing installations. Perhaps some of these facilities will be based in the Bay of Rainbows, which is where China's Moon rover began surveying for minerals in December 2013. Perhaps one of these facilities will be a helium-3 processing plant? Why helium-3? Well, for one thing, it's a substance thought to be far more abundant on the Moon than on Earth. Helium-3 is an isotope of the element that could be used to generate power for thousands of years—assuming fusion reactors become a reality. Everyone knows fossil fuels such as gas and coal will be used up one day but, with one million metric tonnes of helium-3 on the Moon, China's lunar base plans seem commercially viable … as long as someone invents a fusion reactor. But perhaps the Chinese have other things in mind. Perhaps a lunar base is just a stepping stone to bigger and bolder missions, which is exactly what Luan Enjie, a senior adviser to China's lunar program, has told state media. And, to use the Moon as a "springboard" for deep-space exploration, the Chinese will need a lunar base. Can they do it? Sure: they have the technology to do it, they have the buying power to do it, and they are strategically interested in doing it, so chances are they will do it.

As China's Yutu Rover, equipped with its belly-mounted ground-penetrating radar, goes about its business analyzing minerals on the Moon, Bigelow continues to warn those who will listen that a land grab will not only set the Chinese up for mineral rights ownership, but also establish for China a foothold for dominating the space race, not only with regard to the Moon but also future expeditions to Mars. Bigelow hopes that highlighting the possibility of Chinese dominance may alter the American space race lethargy—a position markedly at odds with the national effort during the 1960s that placed a man on the Moon. The problem is that NASA has been placed in a funding stasis for far too long, the Shuttle has been mothballed, American astronauts hitch rides into space on board the Soyuz, and most people in the street are more preoccupied with the latest Miley Cyrus or Justin Bieber escapade than taikonauts taking over the Moon. So, a Chinese takeover of the Moon may well be the greatest threat to America's economic and political future, but nobody really cares, and political reality suggests much of China's competition may not come from other nations, but from private companies such as … Bigelow Aerospace! After all, such competition has precedent. During the Age of Exploration (15th to early 17th centuries), governments, individuals, and trading companies sponsored exploratory missions, and expeditions set up colonies and outposts all over the world. Those who operated the claimed territories did so without interference and enjoyed the spoils of ownership and development. So, if history is an adjudicator as to what China might do once it has set up shop on the Moon, any treaty forbidding ownership of the Moon will most likely have little effect: the Chinese will simply make an unopposed proprietary claim through presence, development, and usage. Or perhaps Bigelow will take a page out of Harriman's book and get there first? How long does he have to land his modules on the lunar surface? Bigelow reckons the Chinese could start laying claims to the lunar surface as early as 2022. That doesn't give him much time. So he continues to emphasize the possibility that the US could be losing the race to own the Moon, hoping the realization may produce the fear factor necessary to motivate the US into action. He also notes that competition with China is not the only option: collaboration might work—after all, a piece of something is better than a piece of nothing and, if that doesn't work, there is always Mars.

Another reason for the US to get to the Moon first

In December 2013, a report in the pro-regime *China Times* announced that China's launch of the Long March-3B rocket earlier that month was part of a long-term plan to turn the Moon into a weapons platform from which the People's Liberation Army (PLA) could launch missiles against terrestrial targets. It may sound like something out of a Michael Bay movie, but this is what Communist Party officials discussed following China's launch of their lunar rover. These officials suggested how the Moon could be transformed into a deadly weapon, used as a military battle station to launch ballistic missiles against military targets on Earth. It was a report that reminded the West that some of Beijing's most jingoistic and aggressive rhetoric is often hidden in plain view, with Chinese military planners willing to go on the record and brag about their agenda to turn China into a forceful military superpower.

So how viable is a Bigelow lunar base and how would such an outpost be realized? Well, we can get some idea by perusing NASA's Space Act Agreement (SAA) with Bigelow Aerospace. Under the agreement, Bigelow will work with commercial space companies to assess and develop options for investments to create infrastructure to support governmental exploration activities alongside revenue-generating private-sector enterprises. The agreement includes a two-phased approach designed to help NASA assess opportunities for collaboration. During the first phase, Bigelow will leverage its relationships with other commercial space companies and its expertise from continuing operations in space to form common objectives between the private sector and NASA. In the second phase, Bigelow will create options for public–private collaboration that lowers costs and takes advantage of rapid implementation. In a way, the SAA is a government contract that gives a green light to a program to establish a lunar base. It's an adventurous deal to be sure, and it's perhaps the only way the US will return to the Moon. As NASA Administrator, Charlie Bolden, has stated: "NASA will not take the lead on a human lunar mission … NASA is not going to the Moon with a human as a primary project probably in my lifetime. And the reason is, we can only do so many things." Which is why NASA has not ruled out the possibility that the agency might play a role in missions led by other countries … or private firms. With the SAA, NASA has picked Bigelow Aerospace to be a lynchpin of this new strategy—a marriage of American knowhow, practical business goals, and good, old-fashioned adventure.

LUNAR BASE

The plans for Bigelow's lunar base were detailed in United States Patent Number US 7,469,864, filed on February 28th, 2008. The title of the patent is "A method for assembling and landing a habitable module on an extraterrestrial mass". Here, we take a look at how this invention may one day see inflatable habitats land on the Moon.

The traditional approach to establishing a lunar base is to launch a number of smaller rigid shelled modules to the Moon and assemble the modules into a larger structure (Figure 8.4) on the surface. Sounds like a reasonable plan, but it is one fraught with shortcomings. For one thing, such an approach is expensive and, for another, it is time-consuming, which increases the chances of failure. Putting all those modules together also means astronauts have to spend an awful long time on the surface. While on the surface, these astronaut construction workers are not only being bombarded by a lethal cocktail of radiation, but are also at risk of their spacesuits being punctured by micrometeorites, the consequences of which are almost certain death. No doubt about it, this architecture is a risky proposition. Rather than using this plan, Bigelow envisions using various elements identified in the Gate 2 Report and landing the whole base on the lunar surface. To achieve this, he plans flying the BA-330-MDS to the Moon with two Docking Node Transporters docked to each side of the module. Added to these two tugs would be four Standard Transit Tugs, which would be used to travel from LEO to lunar orbit. From lunar orbit, the BA-330-MDS would simply land on the Moon with the two Docking Node Transporters on each side. Elegant. Simple and elegant.

Let's describe the nuts and bolts of the operation here. To begin with, a BA-330-MDS is placed in lunar orbit with at least one central node. A second BA-330-MDS is then placed into lunar orbit together with propulsion buses, connecting nodes, and landing pads. The lunar base is then constructed in lunar orbit by inflating the inflatable modules

8.4 Modular lunar base. Courtesy: NASA

and connecting them to the central node. Once that is done, the structure is landed on the Moon by remotely controlling the propulsion buses and landing pads. Sounds simple, doesn't it? Let's go through the process stage by stage.

We'll begin with the BA-330-MDSs in orbit above the lunar surface. Attached to the modules are a number of connecting nodes and one connecting node that acts as a central node that is not initially attached to a module, but to a propulsion bus. The function of the propulsion buses serves to move the spacecraft (soon to become a base) in orbit and from orbit to the surface. The configuration can easily be changed simply by altering the number of modules, propulsion buses, and nodes. Need a four-module lunar base? No problem: just add another module. Of course you can't just keep adding modules without affecting the stability of the structure. After all, you have to be able to land this thing on the lunar surface, so there are limits. Once the modules are joined at the central node, the assembly of the base in orbit is accomplished in a number of ways. One way is to first attach the modules to the central node, after which a node is connected to each propulsion bus and then to a module. Landing pads are then attached to the space buses.

Once the modules are attached to the central node, the assembly is considered to be in the pre-landing configuration. As the habitat/base lands, retro-propulsion kicks in, slowing the descent until the base is landed on the surface. One of the advantages of the inflatable module being inflatable is that the structure can act as a sort of tire and absorb some of the landing force. Of course, such an approach for landing a base isn't restricted to the Moon. The surface can just as easily be an asteroid or even Mars, but let's take a look at the asteroid mission first.

ASTEROID MISSION

Outlining his vision of the future of manned spaceflight during a trip to Kennedy Space Center on April 15th, 2010, President Obama announced that NASA would begin by sending astronauts to an asteroid (Figure 8.5) or near Earth object (NEO) by 2025. The goal was later confirmed in President Obama's US National Space Policy on June 28th, 2010. Why a NEO? Well, a manned mission to a NEO presents new challenges to human space exploration beyond LEO and isn't as risky—or as expensive—as a manned mission to Mars. As well as being a logical next step to deep-space exploration, sending astronauts to NEOs will require a complex system architecture without the development of a costly lander (most asteroids are too small to land on anyway), as would be the case for a Mars mission. Also, missions to NEOs could eventually be extended to simulate a long-duration trip to Mars.

How will NASA get there? Well, first of all, target asteroids must be evaluated, mission lengths must be tested, and multiple system architectures must be analyzed, but the likely vehicle to take astronauts to a LEO will be the Orion (Table 8.1), launched by the Space Launch System (SLS). The current design consists of a crewed capsule attached to a service module, similar to the Apollo design. The capsule housing the astronauts also serves

8.5 Asteroid mission. Courtesy: NASA

Table 8.1. Orion characteristics and performance capabilities.

Endurance	180 days	Crew	Three
Mass to orbit	23,848 kilograms	Diameter	5.03 meters
Habitable volume	8.95 cubic meters		

as the re-entry vehicle when returning to Earth, while the service module provides life support, propulsion, and power for the capsule. Solar arrays on the service module provide electrical power for the Orion.

The table is included to draw your attention to the habitable volume. That's not a lot of space for what might be a fairly long mission (it will depend on which asteroid NASA decide to capture). That's not to say inflatable technology isn't included in the NEO scenario. Once the vehicle makes its rendezvous with the asteroid, the vehicle would engulf the asteroid in a huge inflatable cylinder. But why not attach a Bigelow Expandable Activity Module (BEAM) to increase living space and provide a little more comfort for the astronauts, to say nothing of the increased radiation shielding provided by the Kevlar–Mylar fabric of the habitat? Adding a BEAM would also give the crew better protection from space debris, and all this could be achieved without increasing the system mass significantly because, as we know, these BEAMs don't weigh very much. Adding an inflatable module is something that NASA has thought about, which is why it has been testing its own inflatable test articles at the agency's White Sands Test Facility (WSTF).

MARS MISSION

Dennis Tito wants to send humans to Mars. In 2013, the world's first space tourist announced details of a mission that would see two astronauts set a course for the Red Planet in January 2018, perform a flyby, and return to Earth in 501 days. To help drum up financial support, Tito has founded a non-profit group he calls Inspiration Mars (Figure 8.6). The flight path of the mission's spacecraft will resemble a very large orbit of Earth that leaves and loops back in something mission planners call a *free-return* mission: no critical propulsive maneuvers, no docking, no spacewalks, and no landing required. All the astronauts have to do is keep the life-support systems working. It's an austere and barebones mission.

To achieve this ambitious mission, the self-made millionaire, who paid for his own trip to the International Space Station (ISS) in 2001, has created the non-profit Inspiration Mars Foundation. By sending humans to Mars, Tito hopes the mission will be an inspiration (hence the mission's name) for those back on Earth and will also address one of the biggest challenges facing human space exploration: keeping humans alive and productive in deep space. The mission will use a rare alignment of Earth and Mars to send a spacecraft on a free-return trajectory, essentially flying a manned boomerang around Mars and returning home without needing any major propulsion to get back. The trajectory means there will be no way to abort the mission once the crew departs LEO, leaving the astronauts with only their ingenuity to fix troubles along the way. If all goes to plan, the crew will lift off on January 5th, 2018, on a Delta 4-Heavy or perhaps on a Falcon Heavy. All sorts of

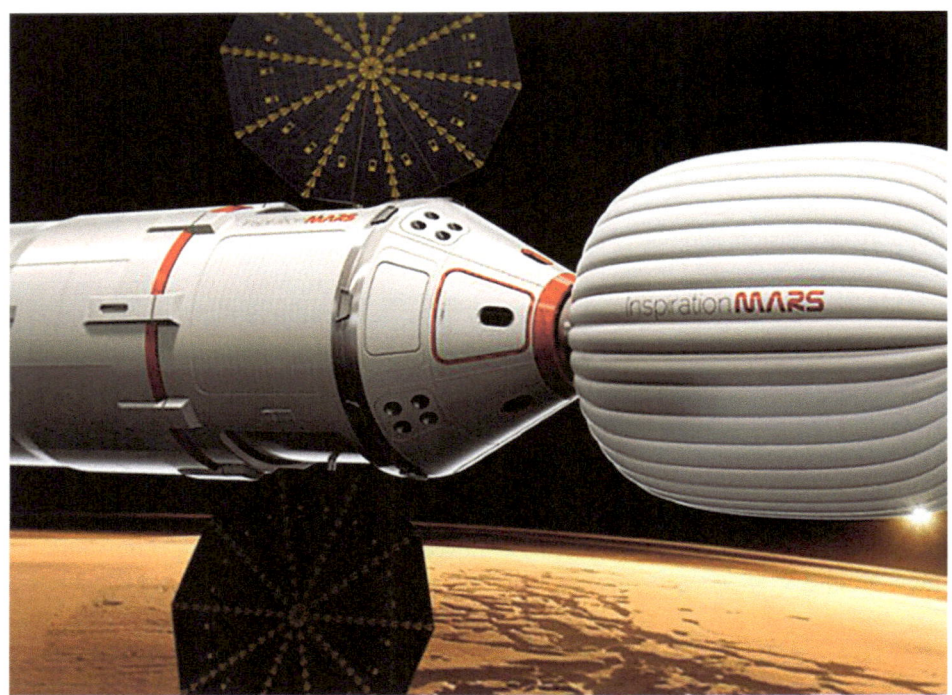

8.6 Inspiration Mars. Courtesy: Inspiration Mars/Wikimedia

challenges will ensue: worries of hardware breakdowns, recycling urine again and again, eating dehydrated food for 501 days straight, the worry of returning to Earth at 14 kilometers per second,[1] and then there is the psychological impact of being cooped up in a cramped craft for months, which brings us to the subject of the inflatable module you can see in Figure 8.6. Will it be a Bigelow inflatable? A Thales-Alenia inflatable? A modified BA-330 DSP perhaps? We don't know, because Inspiration Mars hasn't identified which company will be building the inflatable, although it did say the mission would include a Canadian-made inflatable habitat, which leads us to Thin Red Line, because they're the only Canadian company in the business of building them.

What has this got to do with Bigelow, you may ask? Well, two things. First, Thin Red Line designed and built the pressure-restraining hulls for *Genesis I* and *II*. These units, launched in 2006 and 2007, are now orbiting Earth, and serve as technology

---

[1]   The re-entry velocity is 14.18 kilometers per second, which is higher than any mission that has flown so far. One way to reduce the risk is to break down the entry into two segments, with a 10-day Earth orbit after aerocapture, which would reduce re-entry. Still, even with this approach, a miss no more than a few kilometers in either direction would result in a bad day for the crew, fried or thrown back out into interplanetary space.

demonstrators for Bigelow's space station and lunar base. Second, one of the reasons Bigelow is so secretive is because he doesn't want his competition to gain an advantage in the technology stakes. Admittedly, the competition is limited since there are only two other companies in the business of building inflatable habitats for use in LEO, but Thin Red Line is one of them, although they're not planning on launching space stations or landing bases on the Moon.

So the chances are the first manned Mars mission may not make use of Bigelow's inflatable modules, but what about the reality of that lunar base? According to the Gate 1 and Gate 2 Reports, Bigelow has stated that the Moon is an important objective for his company. In addition to the elegant mission architecture of landing a base on the Moon, the BA-330 modules can easily be modified to include a floor (using Velcro anchors) to provide crew with a surface. In another step towards realizing the landing of that base, Bigelow has plans to test a scaled lunar module (called "the Guide") on Earth. Once initial tests on the Guide are complete, Bigelow will continue testing, eventually migrating to a full-scale system (Bigelow has never been a fan of computer modeling) to provide as much validation as possible. So, while NASA focuses its attention of asteroids and perhaps deep space, commercial companies like Bigelow will be given free rein to pursue lunar options.

Epilog

E. Seedhouse, *Bigelow Aerospace: Colonizing Space One Module at a Time*,
Springer Praxis Books, DOI 10.1007/978-3-319-05197-0

After making his name in real-estate development, Robert Bigelow founded Bigelow Aerospace with the purpose of revolutionizing space commerce via the development of affordable, reliable, and robust expandable space habitats for space agencies and corporate clients. That was in 1999, and the company has been making headlines ever since. Its *Genesis* prototypes are still in orbit, and a 2013 contract with NASA will result in the launch of a BEAM to the ISS in 2015. Still under development is Bigelow's fleet of BA-330 expandable modules, envisioned to function as a commercial space station. The tech is elegant and exciting, and when was the last time you could say that about a NASA project (Mars rovers excepted)? And even when a NASA project was exciting, more often than not it got cancelled. At least with Bigelow, you don't have to worry about the CEO being changed! And those space stations? Link a few of those together and you have an orbital village. It may sound like a concept that is in the realm of science fiction, but it's not—it's happening today. Some may paint Bigelow as just one in a long line of dreamers looking to make money in the nascent commercial orbital economy. After all, dozens of aerospace companies with similar notions have come and gone over the years, with little more to show for their ambition than filing cabinets full of feasibility studies, government grant applications, and glitzy graphics of futuristic hardware. But Bigelow has something none of those predecessors had: a functioning prototype in orbit. As space entrepreneurship develops into something more than just a sci-fi geek's fantasy, this pioneering Las Vegas operation with its revolutionary tech is emerging as one to watch. And with popular interest stoked by SpaceX's commercial cargo flights to the ISS, the idea of commercializing human spaceflight seems here to stay. While others are hammering out the details of how to launch cargo and wealthy tourists into orbit, Bigelow is thinking further over the horizon, to the day when spacefarers—his clients—will want to spend some time up there. And he's focusing on organizations that will need orbiting workspace to produce the kind of zero-gravity research that decades of small-scale experiments on space stations have only been able to hint at.

Many decades ago, Wernher von Braun, the father of American manned spaceflight, dreamt of orbiting space stations, missions to Mars, and lunar bases. In 2014, Robert Bigelow is trying to realize those goals and he's not just dreaming. In many ways, the soul of von Braun's ambitions lives on in the heart of Bigelow Aerospace's orbital and beyond-orbit ambitions. Will he succeed? In most people's minds, Bigelow just might be smart enough, rich enough, and driven enough to pull it off. Either way, you have to admire the man's audacity. I for one hope Bigelow can carry von Braun's torch to LEO and beyond.

Appendix I

Commercial Crew Programs: A Primer

Sierra Nevada Corporation's Dream Chaser vehicle. Courtesy: Sierra Nevada/NASA

After years of keeping orbital transport for astronauts in-house, NASA decided that commercial companies could develop and operate such a system more efficiently and affordably than a government bureaucracy, so Commercial Orbital Transportation Services (COTS) was born. COTS was a NASA program to coordinate the delivery of crew and cargo to the International Space Station (ISS) by private companies. The program was announced on January 18th, 2006, and flew all cargo demonstration flights by September

© Springer International Publishing Switzerland 2015
E. Seedhouse, *Bigelow Aerospace: Colonizing Space One Module at a Time*,
Springer Praxis Books, DOI 10.1007/978-3-319-05197-0

2013. COTS is different from the related Commercial Resupply Services (CRS) program because COTS deals with the development of the vehicles, while CRS deals with the deliveries. Also, COTS does not involve binding contracts whereas CRS does, which means suppliers are liable if they fail to perform. A related program is the Commercial Crew Development (CCDev), designed to develop crew-rotation services. All three programs are managed by NASA's Commercial Crew and Cargo Program Office (C3PO). Here's a little more detail.

COMMERCIAL RESUPPLY SERVICES

- Contracts signed by NASA for delivery of cargo to ISS by commercial firms
- Contracts include a minimum of 12 missions for SpaceX and 8 for Orbital Sciences

The development of the CRS program began in 2006 with the purpose of creating American commercially operated uncrewed cargo vehicles to service the ISS. Development of these cargo-carrying vehicles was under a fixed-price milestone-based program, which meant each company receiving funding had a list of milestones with a dollar value attached to them: companies only received the funding if they achieved the milestones. The first of these CRS contracts was awarded on December 23rd, 2008, when NASA awarded contracts to SpaceX and Orbital Sciences Corporation (usually referred to as just "Orbital"). Under the contracts, SpaceX was to use its Falcon 9 rocket and Dragon spacecraft to haul 20 tonnes of cargo to the ISS while Orbital would use its Antares rocket and Cygnus spacecraft. The contracts, worth a combined US$3.5 billion (US$1.6 billion to SpaceX and US$1.9 billion to Orbital) through 2016, were awarded based on the likelihood of rocket availability and the superior management structures and technical abilities demonstrated by the two companies' proposals.

COMMERCIAL CREW DEVELOPMENT

- Multiphase space technology development program, funded by the US government and administered by NASA
- Run by the Commercial Crew and Cargo Program Office (C3PO)

This is a multiphase space technology development program administered by NASA. The intent of the CCDev program is to stimulate development of privately operated crew vehicles to low Earth orbit (LEO). Under the program, at least two providers will be chosen to deliver crew to the ISS, hopefully no later than 2017. Unlike traditional space industry contractor funding used on the Shuttle, Apollo, Gemini, and Mercury programs, contract funding for the CCDev program contracts is explicitly designed to fund only specific subsystem technology development objectives that NASA wants for NASA purposes; all other system technology development is funded by the commercial contractor.

CCDev-1 Phase 1 (2010–2011)

In the program's first phase (CCDev-1), NASA provided US$50 million during 2010 to five companies, intended to foster research and development into human spaceflight concepts and technologies in the private sector. Later that year, a second set of CCDev proposals were solicited by NASA for technology development project durations of up to 14 months. The proposals selected included Blue Origin, which was awarded US$3.7 million to develop an innovative "pusher" launch-abort system (LAS) and composite pressure vessels. Boeing received US$18 million for development of its Crew Space Transportation (CST)-100 vehicle and Paragon Space Development Corporation was awarded US$1.4 million to develop an environmental control and life-support system (ECLSS) Air Revitalization System (ARS) Engineering Development Unit designed to be used on different commercial crew vehicles. Sierra Nevada Corporation (SNC) received US$20 million for development of Dream Chaser, its reusable spaceplane, capable of transporting cargo and crew to LEO. The fifth company, United Launch Alliance (ULA), received US$6.7 million for an Emergency Detection System (EDS) for human-rating its Evolved Expendable Launch Vehicles (EELVs).

CCDev-2 Phase 2 (2011–2012)

On April 18th, 2011, NASA announced it would award up to nearly US$270 million to four companies as they met CCDev-2 objectives. These objectives included the capability of a vehicle to deliver and return four crewmembers and their equipment, provide crew return in the event of an emergency, serve as a 24-hour safe haven in the event of an emergency, and remain docked to the ISS for 210 days. Winners of funding in the second round of the CCDev Program included Blue Origin, which was awarded US$22 million to develop advanced technologies in support of its orbital vehicle, including launch-abort systems and restartable hydrolox (liquid hydrogen/liquid oxygen) engines, and SNC received US$80 million to develop phase 2 extensions of its lifting-body-inspired Dream Chaser spaceplane. SpaceX was awarded US$75 million to develop an integrated launch-abort system design for its Dragon spacecraft. The system, reputed to have advantages over the more traditional tractor tower approaches used on prior manned space capsules, would be part of the company's Draco maneuvering system, currently used on the Dragon capsule for in-orbit maneuvering and de-orbit burns. Industry juggernaut, Boeing, proposed additional development for their seven-person CST-100 spacecraft, beyond the objectives for the US$18 million received from NASA in CCDev-1. Designed to be used up to 10 times, the capsule would have personnel and cargo configurations, and is designed to be launched by different rockets.

In common with many government-funding processes, to the outsider, NASA's rationale for CCDev awards seemed difficult to grasp. Consider the following proposals. One company proposed continuing work on a project that NASA had already funded to human-rate a pair of highly reliable rockets that at least three companies wanted to use to launch their commercial spacecraft. Another company requested funds to build a new booster that had never flown and which no one intended to use! Which one did NASA fund? Neither. Puzzled? Welcome to the perplexing world of NASA contract awards. The aforementioned companies were ATK, which proposed its Liberty rocket, and ULA, which is developing technologies to human-rate its Atlas V and Delta IV launchers. Why didn't they receive funding? Surprisingly, at least part of the reason seemed to have had little to do with the quality of the proposals. The rationale went something like this: spacecraft proposals were weighted higher than for launch vehicles for the simple reason that American companies have considerable experience developing launch vehicles but no US companies have successfully developed a crew-carrying spacecraft—at least not in the last 30 years. Given this emphasis, it is easier to understand why Boeing and SNC received funds to develop their human vehicles and why SpaceX received funding to human-rate its Dragon spacecraft and Falcon 9 rocket, the development of which NASA has been funding under the COTS program. Blue Origin is also developing a human-rated vehicle, so it received money to develop its biconic capsule and reusable rocket. Meanwhile, proposals that focused solely on rocket development received nothing.

While the experience developing launch vehicles versus the experience developing spacecraft provided part of the justification when it came to awarding funding, the strength and weaknesses of each program also factored into the decision-making, so it's worthwhile taking a look at how NASA makes these assessments. Two factors NASA is particularly interested in are the company's Technical Approach and Business Information. These factors are color-coded, with green indicating a High Level of Confidence and white indicating a Moderate Level of Confidence. For example, ULA, which had sought US$40 million in CCDev-2 funding, rated surprisingly low. NASA's reviewers identified several strengths in ULA's proposal, among them the company's use of existing flight-proven vehicles and infrastructure, its adaptable emergency detection system, a strong performance capability for crew-abort scenarios, and an effective and integrated organizational structure. Its business information didn't fare too badly either, since it was deemed suitable to deliver proposed capabilities, it had a strong, highly experienced management team, possessed the requisite facilities, and had experienced and knowledgeable suppliers. With such a strong résumé, you may be wondering why the company wasn't funded, but the reviewers also found weaknesses, among them a lack of definition of critical path to an initial launch capability and correlation to CCDev-2 efforts, and a failure to adequately describe the commercial market to which it would provide products and services.

Ultimately, NASA deemed that ULA's work on their existing launch vehicles was not on the critical path for any crew transportation system and therefore the company did not accelerate the availability of crew transportation capabilities which, after all, was a primary goal of the funding announcement. Nevertheless, while this assessment was true, it seemed strange that NASA would deny funding to a company that is developing a rocket that two CCDev-2-funded companies hoped to use: SNC and Blue Origin, which both received CCDev-2 funding, had stated they wanted to use ULA's Atlas V rocket to launch their crew vehicles (Blue Origin eventually shifted to its own reusable launcher). Boeing, which also received CCDev-2 funds, had also expressed their intent of using Atlas V to launch its CST-100 spacecraft, although the vehicle is being designed for multiple launchers.

At first glance, another strange case was the non-funding of ATK's Liberty rocket. This launcher comprises a first stage derived from the canceled Shuttle-derived Ares I booster and the second stage of Europe's Ariane 5. From the NASA reviewer's perspective, while these technologies had excellent flight heritages, the problem was that no one had committed to flying on the rocket. You can understand the agency's concern: NASA could fund the Liberty all the way through the development phase but there would always be the risk that no spacecraft developer would select the launch vehicle as part of its design!

Another black mark against ATK was their failure to provide NASA with sufficient details to assess launch vehicle environments (such as staging and abort scenarios) on the company's proposed upper stage or at the crewed spacecraft interface. While the company provided a solid technical approach, their details on environments didn't provide NASA with enough confidence in accelerating this launch vehicle for use with the variety of different crew spacecraft. So, rather than use limited CCDev-2 funds on a launch vehicle with a questionable technical approach, the agency instead decided to select an extra spacecraft.

Commercial Crew Integrated Capability Phase 3 (2012–2014)

On August 3rd, 2012, NASA announced awards made to three American commercial companies under yet another funding program: CCiCap. Advances made by these companies under the signed Space Act Agreements (SAAs) through the agency's CCiCap initiative are intended to lead to the availability of commercial human spaceflight services for government and commercial customers. The CCiCap partners were SNC, which received US$212.5 million; SpaceX, which received US$440 million; and the Boeing Company, which received US$460 million.

As an initiative of NASA's Commercial Crew Program (CCP), CCiCap is an administration priority. The objective of the CCP is to facilitate the development of a US commercial crew space transportation capability with the goal of achieving safe, reliable, and cost-effective access to and from the ISS and LEO. After the capability is matured and expected to be available to the government and other customers, NASA plans to contract to purchase commercial services to meet its station crew transportation needs.

The new CCiCAP agreements follow two previous initiatives by NASA to spur the development of transportation subsystems, and represent the next phase of US commercial human space transportation, in which industry partners develop crew transportation capabilities as fully integrated systems. Between now and May 31st, 2014, NASA's partners will perform tests and mature integrated designs. This will then set the stage for a future activity that will launch crewed orbital demonstration missions to LEO by the end of the decade.

Appendix II

Commercial Spaceflight Training

A NASA astronaut prepares for mission training in the Sonny Carter Training Facility, Neutral Buoyancy Laboratory, at Johnson Space Center. Courtesy: NASA

WAYPOINT 2 SPACE

If Bigelow's clients (budding spaceflight participants, or SFPs) are trained by Waypoint 2 Space, they will most likely be enrolled in the company's orbital training program, which contains three sub-programs/modules, as follows.

© Springer International Publishing Switzerland 2015
E. Seedhouse, *Bigelow Aerospace: Colonizing Space One Module at a Time*,
Springer Praxis Books, DOI 10.1007/978-3-319-05197-0

1. Orbital Launch Vehicle Operations

In this module, SFPs will develop a knowledge of emergency ingress and egress, launch and flight procedures, life-support operations, and emergency procedures. SFPs will also be briefed on spaceflight safety and risks encountered during orbital flight. The course includes the following modules:

Spaceflight Physiology 1 reviews the physiological effects that occur during spaceflight. The module includes the following sub-modules:

- human anatomy specific to the SFP during orbital flight;
- basic gas laws;
- physiological divisions of the atmosphere;
- altitude physiology and mitigation measures;
- G-forces, self-imposed stress, motion sickness, and mitigation;
- potential environmental stressors;
- incapacitation and health maintenance prior to spaceflight.

Orbital Spaceflight Environment 1 reviews spaceflight dynamics to prepare the SFP for spaceflight. The module includes the following sub-modules:

- Earth's atmospheric structure and radiation in low Earth orbit;
- orbital mechanics principles;
- vehicle life support, communications, propulsion, and attitude control;
- operator's mission architecture and Mission Control roles and responsibilities.

Orbital Spaceflight Environment 2: Orbital Transitions and Close Vehicle Proximity Operations introduces SFPs to vehicle rendezvous. The module includes the following sub-modules:

- orbital mechanics, including Keplerian orbital elements and changes in orbital altitude and inclination;
- introduction to vehicle rendezvous, methods of approach, vehicle docking, and close-proximity operations;
- introduction to orbit phasing, spacecraft true anomaly, velocity matching, and station-keeping;
- mathematical equations required to calculate proper rendezvous solutions.

Crew Resource Management Review introduces human factors issues present during flight and tools for working in a crew environment. The module includes the following sub-modules:

- interpersonal communication;
- conflict and stress management, situational awareness;
- decision-making and professionalism.

Pressure Suit Review reviews pressure suit fit, form and function, and fitting and operation of the pressure suit used by the operator. The module reviews:

- terminology of pressure suit hardware;
- donning and doffing procedures, suit operations, and suit safety precaution.

Spaceflight Safety and Risk Briefing briefs the SFP in accordance with Federal Aviation Administration (FAA) requirements on the physical risks and hazards associated with spaceflight and the safety record of the launch vehicle type. The module includes the following topics:

- the number of people who have flown an orbital flight;
- the number of people who have died during orbital flight;
- the number of launch and re-entries conducted with people on board and any catastrophic failures associated with these flights;
- the operator's safety record post verification of the vehicle to include any incidents or mishaps during any phase of flight;
- corrective actions taken by the operator to resolve incidents or mishaps;
- rights afforded to the SFP prior to flight.

Launch Vehicle Operations 1: Nominal Procedures Training introduces the SFP to the nominal procedures used by the operator during orbital launch vehicle operations utilizing a mock-up. During the course, the SFP will:

- become familiar with flight hardware;
- demonstrate vehicle ingress and egress in accordance with the operator's procedures;
- demonstrate operation of the SFP restraint system;
- become knowledgeable on vehicle life-support systems, nominal launch, and orbital insertion procedures;
- be familiar with microgravity translation points and restrictions imposed by the operator;
- be provided with training leading to mastery of systems and equipment operations.

Launch Vehicle Operations 2: Emergency Procedures Training and Simulation trains the SFP on emergency procedures. During the course, the SFP will:

- be introduced to emergency procedures;
- be trained to respond to a cabin fire, cabin smoke, loss of cabin pressure, and techniques associated with emergency ground egress;
- develop a working knowledge of the vehicle operator's emergency egress procedures, life support, and survival equipment;
- perform simulated emergency drills;
- be provided with training leading to mastery of systems and equipment operations.

Launch Vehicle Systems 3: Docking Systems trains the SFP on an operator's launch vehicle docking system, equipment, pre-docking, and post-departure procedures utilizing a mock-up in preparation for flight. During this module, the SFP will:

- be introduced to flight hardware and docking system components, controls, and equipment;
- become knowledgeable on pre-docking and post-departure procedures, SFP roles;
- demonstrate proper operation of the docking system and equipment;
- be provided with training leading to mastery of systems and equipment operations.

Launch Vehicle Systems 4: Galley, Health and Hygiene System trains the SFP on an operator's orbital vehicle galley, health and hygiene systems, and equipment utilizing a mock-up in preparation for flight. During this course, the SFP will:

- become familiar with flight hardware, galley, health and hygiene system components, controls, and equipment;
- demonstrate proper operation of the galley, health, and hygiene systems and equipment;
- be provided with training leading to mastery of systems and equipment operations.

Launch Vehicle Systems 5: Launch System Architecture introduces the SFP to an operator's launch vehicle mission support structure.

Launch Vehicle Systems 6: Payload and Specialized Systems introduces the SFP to unique payloads or systems installed on the operator's orbital vehicle utilizing mock-ups. During this course, the SFP will:

- be introduced to payload flight hardware and/or specialized system components, controls, and equipment;
- demonstrate proper operation of the payload and/or specialized systems and equipment;
- be provided with training leading to mastery of systems and equipment operations.

2. IVA Operations

This course will advance the SFP's knowledge in extended-duration spaceflight physiology and the spaceflight environment to include orbital transitions and vehicle close-proximity operations through academic instruction. Intravehicular activity (IVA) ground training events will introduce the SFP to operations occurring during a stay of 10 days or more. The course includes the following modules:

Spaceflight Physiology 2: Extended Spaceflight Effects introduces the SFP to physiological issues experienced during extended-duration spaceflight. Topics include:

- how and why the body reacts during extended spaceflight;
- musculoskeletal system deconditioning, fluid redistribution, and reduction of aerobic capacity;

- effects of microgravity on the cardiovascular systems;
- psychological risks, radiation exposure, fatigue, and disruption of sleep patterns;
- methods of mitigating physiological decay;
- negative psychological effects as a result of extended microgravity exposure.

Spaceflight Physiology 3: Post-Flight Effects introduces the SFP to the physiological effects on the body when returning to Earth after extended exposure to microgravity. Topics include:

- how and why the body reacts upon return from extended spaceflight;
- loads exerted on the musculoskeletal system, orthostatic intolerance, and orthostatic hypotension;
- psychological risks such as asthenization syndrome;
- methods of mitigating physiological and psychological effects when returning to Earth.

Orbital Spaceflight Environment 3: Low Earth Orbit familiarizes the SFP with the low Earth orbit environment. Topics include:

- low Earth orbit spatial boundaries;
- requirement, advantages, and limitations of reaching and maintaining low Earth orbit;
- orbital periods, vehicle velocity, and temperature extremes while in low Earth orbit;
- re-entry, mission architecture, radiation, and micrometeorite/orbital debris hazards and mitigation measures executed during low Earth orbit.

Orbital Vehicle Systems 1: Vehicle Orientation provides the SFP with an orientation to the orbital vehicle. The SFP will become familiar with:

- the general layout of the orbital vehicle habitat;
- system interface locations and personal use stations;
- living quarters, life-support equipment, and personal protection equipment.

Orbital Vehicle Systems 2: Life Support Systems trains the SFP on an operator's life-support systems and equipment utilizing a mock-up. During the course, the SFP will:

- become familiar with flight hardware and life-support system components and controls;
- demonstrate proper operation of the life-support system and equipment;
- be provided with training leading to mastery of systems and equipment operations.

Orbital Vehicle Systems 3: Environmental Systems trains the SFP on an operator's vehicle environmental system and equipment utilizing a mock-up. During the course, the SFP will:

- become familiar with flight hardware and environmental system components and controls;
- demonstrate proper operation of the environmental system and equipment;
- be provided with training leading to mastery of systems and equipment operations.

Orbital Vehicle Systems 4: Galley, Health, Hygiene Systems and Living Quarters trains the SFP on the operator's orbital vehicle galley, health and hygiene systems, and personal living space equipment utilizing a mock-up. During the course, the SFP will:

- become knowledgeable on flight hardware and galley, health and hygiene system components, controls, and equipment;
- demonstrate proper operation of the galley, health and hygiene systems, and equipment;
- become knowledgeable on personal living quarters, equipment, and exercise protocols;
- be provided with training leading to mastery of systems and equipment operations.

Orbital Vehicle Systems 5: Docking Systems trains the SFP on the operator's vehicle docking system, equipment, and post-docking and pre-departure procedures utilizing a mock-up. During the course, the SFP will:

- become familiar with flight hardware and docking system components and controls;
- become knowledgeable on and demonstrate SFP roles during post-docking and pre-departure procedures;
- demonstrate proper operation of the docking system and equipment;
- be provided with training leading to mastery of systems and equipment operations.

Orbital Vehicle Systems 6: Payload and Specialized Systems trains the SFP on unique payloads or systems installed on the operator's vehicle utilizing mock-ups. During the course, the SFP will:

- become familiar with payload flight hardware and/or specialized system components, controls, and equipment;
- demonstrate proper operation of payloads, specialized systems, and equipment;
- be provided with training leading to mastery of systems and equipment operations.

Orbital Vehicle Emergency Procedures Training and Simulation trains the SFP on emergency procedures in accordance FAA requirements. During the course, the SFP will:

- be introduced to general and operator's emergency procedures;
- be trained on SFP response to a cabin fire, cabin smoke, loss of cabin pressure, and techniques associated with emergency ground egress;
- develop a working knowledge of the vehicle operator's emergency egress procedures, life support, and survival equipment;
- perform simulated emergency drills;
- be provided with training leading to mastery of systems and equipment operations.

Spaceflight Dynamics and Control Simulation introduces the SFP to working in a crew and Mission Control environment. During the simulation, the SFP will:

- work to complete mission objectives;
- utilize spacecraft attitude control, propulsion, telemetry, and communication systems;

- exercise crew resource management (CRM) principles and techniques manifested through problem solving;
- be provided with training leading to improvements in operations requiring CRM.

Accelerated G-Force Adaptation subjects the SFP to high G-force load effects and unusual attitudes. During training, the SFP will:

- review gas laws, proper mitigation, and self-clearing procedures;
- review hypoxic experiences during the chamber flight;
- understand the accelerated G-force profile and survival equipment used during flight;
- be introduced to countermeasures designed to prevent loss of consciousness resulting from accelerated G-force loading;
- become familiar with slow and rapid G-force onset in the linear, radial, and angular acceleration vectors while exercising physiological G-force countermeasures;
- experience the effect of gas laws while exercising proper mitigation and self-clearing procedures.

Microgravity Adaptation subjects the SFP to microgravity effects coupled by stimulation of the vestibular and visual senses as a result of unusual attitudes. During the course, the SFP will:

- review accelerated G-force flights to include the IMSAFE[1] personal checklist;
- review microgravity translation techniques during pressure suit operations;
- be introduced to periods of microgravity and demonstrate microgravity operations;
- become knowledgeable of physiological effects associated with changes in gravity;
- exercise recovery from unusual attitudes and demonstrate proper translation techniques.

3. EVA Operations

EVA Operations provides the SFP with specialized training focusing on planned EVA tasks and contingent operations. The course includes advanced pressure suit operations, translation techniques, EVA tools and crew aids, procedure execution, and EVA flight rules and protocols. The course includes the following modules:

---

[1] IMSAFE is a mnemonic used by pilots—or SFPs in this case—to assess their fitness to fly:
- Illness: is the SFP suffering from any illness that might affect them in flight?
- Medication: is the SFP taking any prescription or over-the-counter drugs?
- Stress: psychological or emotional factors which might affect the SFP's performance.
- Alcohol: SFPs should consider their alcohol consumption within 8–24 hours prior to flight.
- Fatigue: has the SFP had sufficient sleep and rest?
- Eating: is the SFP sufficiently nourished?

Spaceflight Physiology 4: EVA Physiology introduces the SFP to physiological issues, considerations, hazards, and mitigation to prepare and execute an EVA. The SFP will gain an understanding of:

- spinal elongation, decompression sickness, and physical ailments exhibited during EVA;
- micrometeorite risks, radiation exposure, and fatigue.

Orbital Spaceflight Environment 4: EVA Environment introduces the SFP to the environment experienced during EVA to include hazards and mitigation techniques.

Orbital Spaceflight Environment 5: EVA Operations introduces the SFP to the EVA operational environment to include the operator's mission support architecture, use of procedures, translation techniques, protocols, and flight rules.

EVA Systems 1: EVA Pressure Suit introduces the SFP to EVA pressure suit fit, form, and function preparing the SFP for EVA. During the course, the SFP will:

- be fitted in the EVA pressure suit used by the operator;
- become familiar with the proper nomenclature of EVA pressure suit hardware;
- become familiar with associated ancillary items used for comfort;
- develop a working knowledge of IVA and EVA don and doff procedures, suit operations, and suit safety precautions;
- perform a suit-sizing evaluation to determine any adjustments needed in fit;
- be evaluated for mobility.

EVA Systems 2: Tools and Crew Aids trains the SFP on the operator's EVA tools and crew aid equipment designed and used for EVA utilizing mock-ups and hardware in preparation for and execution of EVA. During the course, the SFP will:

- become familiar with EVA flight hardware, restraint devices, and safety equipment;
- be instructed on and demonstrate operation of the EVA tools and crew aid equipment in accordance with the operator's IVA and EVA procedures;
- be introduced to and become knowledgeable on EVA preparation protocols;
- be provided with training leading to mastery of systems and equipment operations.

EVA Systems 3: Airlock Systems and Operation trains the SFP on the operator's vehicle airlock systems and equipment utilizing a mock-up in preparation for EVA. During the course, the SFP will:

- become knowledgeable on hardware, airlock systems components, controls, and equipment;
- demonstrate operation of the airlock systems and equipment in accordance with the operator's IVA and EVA procedures;
- become knowledgeable on EVA preparation protocols;
- be provided with training leading to mastery of systems and equipment operations.

EVA Systems 4: IVA Nominal Procedures trains the SFP on procedures and protocols used in IVA to prepare for EVA utilizing mock-ups. During the course, the SFP will:

- become knowledgeable on flight hardware and IVA procedures used pre and post EVA;
- demonstrate gathering of EVA hardware from stowed locations and proper operation of airlock systems;
- demonstrate EVA pressure suit preparation, ingress/egress;
- demonstrate pre-EVA physiological protocols, depressurization, and re-pressurization;
- be provided with training leading to mastery of systems and equipment operations.

EVA Systems 4: IVA Emergency Procedures trains the SFP on execution of emergency procedures and protocols used in IVA to respond to EVA emergencies utilizing mock-ups. During the course, the SFP will:

- become knowledgeable on EVA flight hardware and IVA emergency procedures used pre, during, and post EVA;
- demonstrate emergency operation of airlock systems, EVA pressure suit emergency egress, and emergency re-pressurization protocols;
- be provided with training leading to mastery of systems and equipment operations.

EVA Systems 5: EVA Nominal Procedures trains the SFP on execution of nominal EVA procedures. During the course, the SFP will:

- become knowledgeable on planned EVA procedures;
- demonstrate proper translation routes, techniques, and keep-out zones;
- be provided with training leading to mastery of systems and equipment operations.

EVA Systems 6: EVA Emergency Procedures trains the SFP on execution of emergency procedures and protocols used to respond to EVA emergencies utilizing mock-ups. During the course, the SFP will:

- become knowledgeable on flight hardware and EVA emergency procedures;
- demonstrate emergency procedure operations of the EVA pressure suit and emergency vehicle ingress;
- be introduced to and instructed on incapacitated rescue techniques and procedures;
- be provided with training leading to mastery of systems and equipment operations.

EVA Systems 7: EVA Simulation prepares the SFP for the planned EVA utilizing fidelity correct mock-ups in a neutrally buoyant or neutral gravity environment simulating microgravity procedures for orbital EVA or terrestrial based facility if performing lunar EVA. During the course, the SFP will:

- review topics covered during Nominal Procedures to include EVA objectives;
- don and perform an operational check of the operator's EVA pressure suit then enter the neutrally buoyant or neutral gravity environment;

- perform the planned nominal EVA procedure while incorporating emergency procedures and incapacitated rescue;
- exercise CRM principles and techniques focusing on critical communication training;
- be provided with continued practical training leading to mastery of systems and equipment operations.

AMERICAN ASTRONAUTICS INSTITUTE

An alternative for Bigelow's clients is the Flight Operations Engineer Program of the American Astronautics Institute (AAI). This is a graduate-credential level program that integrates astronaut training and provides a comprehensive education for those needing a thorough grounding in commercial space operations. Those enrolled in this program are taught to manage space missions and are trained in a variety of spacecraft systems and science applications. The program combines four months of virtual instruction with a 22-day intensive training program near NASA's Kennedy Space Center. The curriculum comprises the following eight modules:

SO101: Spaceflight Physiology

This course covers the unique aspects of health maintenance of individuals exposed to spaceflight, including an overview of the physiological changes resulting from prolonged exposures to weightlessness and the establishment of countermeasures.

Outline

- Physiological adaptation to spaceflight
- Vestibular system
- Neurophysiology and human performance
- Cardiopulmonary system
- Body mass and neuromuscular function
- Bone and mineral metabolism
- Health maintenance of space crews
- Historical selection criteria of astronauts
- Medical training of space crews
- Deconditioning and countermeasures
- Medical problems in spaceflight
- Toxic hazards in space habitations
- Radiation exposure
- Health maintenance in space

SO102: Spatial Orientation and Motion Sickness in Spaceflight: Training and Simulation

This module is administered at Brandeis University near Boston, MA. The module is based on elements of prior research that are most relevant to the circumstances and G profiles that commercial astronauts are likely to perform.

Outline

The program consists of limited classroom learning, interactive computer exercises, and training in the following facilities:

> *Rotating room/artificial gravity (AG) facility.* This facility is an analog environment used to demonstrate and train for disorientation, movement control, and sensory illusions during high G and G-transitions. In this environment, SFPs are trained to recognize the forces acting on the body and to coordinate movements as needed to avoid errors. Motion sickness awareness training and possible pre-adaptation are also conducted here.

> *Multi-Axis Rotation and Tilt (MART) device.* This device subjects SFPs to oscillations/rotation/tilts in two axes simultaneously to help them recognize and deal with disorientation, as well as perception of balance and balance control.

> *Vection chamber/optokinetic drum.* The striped walls and floor of this cylindrical chamber can rotate independently of each other, causing illusory experiences such as self-motion and/or changing body dimensions or configuration. The device is used to train SFPs to recognize sensory illusions and be aware of their compelling nature.

SO 103: Spaceflight Operations: Training and Simulation

This course provides the SFP with analog mission experience using (1) high-G environments, (2) microgravity environments, and (3) motion simulation environments emphasizing proper procedural execution and proper CRM techniques.

Outline

> *Microgravity analog training.* A specially modified Airbus 300 is used to create periods of research-grade microgravity with durations of approximately 25 seconds each. This aircraft is unique in that the microgravity parabolas are computer-controlled, leading to superior environments best duplicating the space environment. Coursework is divided between two days of training and each day one flight of 25 parabolas is flown. On the first day, the SFP learn skills associated with microgravity adaptation as well as fundamental skills such as translation, instrument operations, and seat ingress and egress. On the second day, the SFP performs typical operations while in a spacesuit, as well as spacesuit don and doff procedures.

Ascent and re-entry analog training. To simulate the ascent and re-entry phases, an Extra 300 aircraft is used to maintain Gz profiles of up to 4 Gs. SFPs are trained to use the Anti-G Straining Maneuver (AGSM) to mitigate G-LOC, or G-induced blackout.

CRM training. A three-axis simulator is used to introduce the SFP to the use of effective CRM techniques.

Comprehensive biometric analysis is provided by AAI's aerospace physiology team during SFP training activities. SFPs receive feedback from an electrocardiogram (ECG) analysis, pulse oximetry, and blood pressure.

SO104: Spacecraft Egress and Rescue Operations

This module provides spaceflight crews with the knowledge and skills necessary to react appropriately to pad emergencies and post-landing contingencies requiring an emergency egress.

Outline

- Planning for nominal and contingency landings
- Nominal rescue operations
- Contingency rescue operations for land-landing spacecraft
- Contingency rescue operations for water-landing spacecraft
- Global search and rescue (SAR) response resources supporting contingency landings
- Contingency rescue and recovery planning
- Post-landing contingencies
- Spacecraft egress
- Pad egress failure environments
- Pad egress design and operations
- Egress system design and operations
- Egress procedures and operations
- Assessing probabilities and effects of injuries and deconditioning
- Rescue operations involving injured crewmembers
- Assessing the effects of deconditioning on egress
- Effects of entrapment on egress
- Emergency post-landing survival kits
- Medical resources
- Fundamental egress operations in commercial manned spaceflight
- Safety and survival equipment utilization and deployment
- Coping with physiological and psychological stress
- Introduction of rescue devices and simulated rescues
- Preparation for emergency landing situations in a spacecraft
- Evacuation through an emergency exit from a spacecraft
- Physics and physiology for use of compressed air

- Pre-flight inspection, egress considerations, and clearing procedures using an emergency breathing device (EBD)
- Conducting an emergency egress on breath hold utilizing the Shallow Water Egress Trainer (SWET)
- Conducting an emergency egress with an EBD utilizing the SWET
- Evacuation and escape training utilizing the Modular Egress Training Simulator (METS™) with and without utilizing an EBD

SO105: Wilderness and Sea Survival for Spaceflight Crews

This course provides essential skills required for spaceflight crews that might be subject to an off-nominal landing where survival skills may be required prior to rescue.

Outline

- General survival concepts
- Anatomy of a survival situation
- Survival psychology and leadership
- Coping with physiological and psychological stress
- Post-landing spacecraft operations
- Survival skills in terrestrial environments
- Shelter
- Fire
- Water
- Food
- Dangerous animals and environments
- Survival skills in marine environments
- Safety and survival equipment utilization and deployment
- Hypothermia mitigation and sea survival
- Personal rescue techniques and use of life rafts
- Characteristics of personal floatation devices
- Life raft deployment, entry, and simulated emergency scenarios
- Individual and group sea surface formations
- SAR resources and equipment
- Priorities in cold environments
- Hypothermia/frostbite awareness
- Generating heat from available resources
- Cold weather shelters and using the spacecraft as shelter
- Food and water in a cold environment
- Priorities in a hot environment
- Shelters from heat
- Water requirements
- Finding or manufacturing water

- Heat illness review
- Facilitating rescue
- When to travel
- Fundamental navigation skills
- GPS and communication
- SAR for remote locations
- Signaling and pyrotechnics
- Injuries and illness
- Spacecraft survival kits
- History and evolution
- Essential equipment
- Using medical equipment and supplies for survival

SO 106: Life Support Systems and Spacesuit Operations

In this module, the SFP learns to recognize the effects of off-nominal environments, such as hypoxia and hyperoxia, and effective use of spacesuits to include human requirements, subsystems, architecture, operation, and maintenance.

Outline

- Spacecraft life-support systems
- The extraterrestrial environment
- Fundamental requirements of life-support systems
- Fundamentals of life-support systems
- Physicochemical life-support systems
- Bioregenerative life-support systems
- Introduction to spacesuits
- Human interface: heating/cooling, oxygen, vision, touch, communications, waste, water, food
- Spacesuit subsystems outline: enclosure, thermal regulation, air flow, communications, micrometeoroid protection, gloves, waste management
- Spacesuit enclosure architecture: outline of layers, pressure joint design, bearings, donning and doffing considerations, helmet design
- Spacesuit operations
- Basic spacesuit operation: donning and doffing
- Basic spacesuit operation: programming, pressure regulation, communications, sizing
- Spacesuit maintenance and checkout: drying and cleaning requirements
- Spacesuit leak checks and repair, visual inspections
- Off-nominal environments
- Hypoxia recognition
- Hyperoxia

SO107: Aerospace Environment

This module provides the SFP with an understanding of the re-entry and orbital environments within which aerospace vehicles operate, and the impacts these environments have on spacecraft crews.

Outline

- Survey of the solar–terrestrial system
- Plasma physics overview
- Solar physics overview
- Terrestrial atmosphere overview
- Components of the solar–terrestrial system
- The Sun
- The solar wind
- The geomagnetic field
- The radiation belts
- The magnetosphere
- The neutral atmosphere
- The ionosphere
- Noctilucent clouds and polar mesospheric summer echoes (PMSE)
- Sprites
- Ionospheric variability and perturbations
- Impact on spacecraft systems
- Spacecraft charging
- Satellite drag
- Debris and impact phenomena
- Radio propagation and communications
- Particles and radiation health hazards
- Radiation hazards to electronic systems and single-event upsets
- Mission planning and safety

SO108: Science Applications in Space Missions

This course is a survey of science applications that have been conducted in space, including imagery and remote sensing, broad scientific uses of microgravity, and space life science applications.

Outline

- Microgravity sciences
- Fluids: "simple" two-phase fluids, complex fluids, phase-change

- Combustion and other reactions
- Material sciences and processes
- Space life sciences
- Imagery and remote sensing
- Visible imagery
- Multispectral imagery
- Laser remote sensing
- Neutral and charged particle detection
- Aerospace medicine

Index

© Springer International Publishing Switzerland 2015
E. Seedhouse, *Bigelow Aerospace: Colonizing Space One Module at a Time*,
Springer Praxis Books, DOI 10.1007/978-3-319-05197-0